Practice Papers for SQA Exams

Intermediate 2 | Units 1, 2 and Applications

Mathematics

01/151009

ISBN 978-1-84372-798-9

Published by
Leckie & Leckie Ltd, 3rd floor, 4 Queen Street, Edinburgh, EH2 1JE
Tel: 0131 220 6831 Fax: 0131 225 9987
enquiries@leckieandleckie.co.uk www.leckieandleckie.co.uk

A CIP Catalogue record for this book is available from the British Library.

Leckie & Leckie Ltd is a division of Huveaux plc.

Introduction

Layout of the Book

This book contains practice exam papers, which mirror the actual SQA exam as much as possible. The layout, paper colour and question level are all similar to the actual exam that you will sit, so that you are familiar with what the exam paper will look like.

The solutions section is at the back of the book. The full worked solution is given to each question so that you can see how the right answer has been arrived at. The solutions are accompanied by a commentary which includes further explanations and advice. There is also an indication of how the marks are allocated and, where relevant, what the examiners will be looking for. Reference is made at times to the relevant sections in Leckie & Leckie's book 'Intermediate 2 Maths Revision Notes'.

Revision advice is provided in this introductory section of the book, so please read on!

How to use This Book

The Practice Papers can be used in two main ways:

1. You can complete an entire practice paper as preparation for the final exam. If you would like to use the book in this way, you can either complete the practice paper under exam style conditions by setting yourself a time for each paper and answering it as well as possible without using any references or notes. Alternatively, you can answer the practice paper questions as a revision exercise, using your notes to produce a model answer. Your teacher may mark these for you.
2. You can use the Topic Index at the front of this book to find all the questions within the book that deal with a specific topic. This allows you to focus specifically on areas that you particularly want to revise or, if you are mid-way through your course, it lets you practise answering exam-style questions for just those topics that you have studied.

Revision Advice

Work out a revision timetable for each week's work in advance – remember to cover all of your subjects and to leave time for homework and breaks. For example:

Day	6pm–6.45pm	7pm–8pm	8.15pm–9pm	9.15pm–10pm
Monday	Homework	Homework	English Revision	Chemistry Revision
Tuesday	Maths Revision	Physics Revision	Homework	Free
Wednesday	Geography Revision	Modern Studies Revision	English Revision	French Revision
Thursday	Homework	Maths Revision	Chemistry Revision	Free
Friday	Geography Revision	French Revision	Free	Free
Saturday	Free	Free	Free	Free
Sunday	Modern Studies Revision	Maths Revision	Modern Studies	Homework

Make sure that you have at least one evening free a week to relax, socialise and re-charge your batteries. It also gives your brain a chance to process the information that you have been feeding it all week.

Arrange your study time into one hour or 30 minutes sessions, with a break between sessions e.g. 6pm–7pm, 7.15pm–7.45pm, 8pm–9pm. Try to start studying as early as possible in the evening when your brain is still alert and be aware that the longer you put off starting, the harder it will be to start!

Study a different subject in each session, except for the day before an exam.

Do something different during your breaks between study sessions – have a cup of tea, or listen to some music. Don't let your 15 minutes expanded into 20 or 25 minutes though!

Have your class notes and any textbooks available for your revision to hand as well as plenty of blank paper, a pen, etc. You should take note of any topic area that you are having particular difficulty with, as and when the difficulty arises. Revisit that question later having revised that topic area by attempting some further questions from the exercises in your textbook.

Revising for a Maths Exam is different from revising for some of your other subjects. Revision is only effective if you are trying to solve problems. You may like to make a list of 'Key Questions' with the dates of your various attempts (successful or not!). These should be questions that you have had real difficulty with.

Key Question	**1st Attempt**		**2nd Attempt**		**3rd Attempt**	
Textbook P56 Q3a	18/2/10	X	21/2/10	√	28/2/10	√
Practice Exam A Paper1 Q5	25/2/10	X	28/2/10	X	3/3/10	
2008 SQA Paper, Paper2 Q4c	27/2/10	X	2/3/10			

The method for working this list is as follows:

1. Any attempt at a question should be dated.

2. A tick or cross should be entered to mark the success or failure of each attempt.

3. A date for your next attempt at that question should be entered:

for an unsuccessful attempt – 3 days later

for a successful attempt – 1 week later

4. After two successful attempts remove that question from the list

(you can assume the question has been learnt!)

Using 'The List' method for revising for your Maths Exam ensures that your revision is focused on the difficulties you have had and that you are actively trying to overcome these difficulties.

Finally forget or ignore all or some of the advice in this section if you are happy with your present way of studying. Everyone revises differently, so find a way that works for you!

Transfer Your Knowledge

As well as using your class notes and textbooks to revise, these practice papers will also be a useful revision tool as they will help you to get used to answering exam style questions. You may find as you work through the questions that you find an example that you haven't come across before. Don't worry! There may be several reasons for this. You may have come across a question on a topic that you have not yet covered in class. Check with your teacher to find out if this is the case. Or it may be the case that the wording or the context of the question is unfamiliar. This is often the case with reasoning questions in the Maths Exam. Once you have familiarised yourself with the worked solutions, in most cases you will find that the question is using mathematical techniques with which you are familiar. In either case you should revisit that question later to check that you can successfully solve it.

Trigger Words

In the practice papers and in the exam itself, a number of 'trigger words' will be used in the questions. These trigger words should help you identify a process or a technique that is expected in your solution to that part of the question. If you familiarise yourself with these trigger words, it will help you to structure your solutions more effectively.

Trigger Word	Meaning/ Explanation
Evaluate	Carry out a calculation to give an answer that is a value.
Hence	You must use the result of the previous part of the question to complete your solution. No marks will be given if you use an alternative method that does not use the previous answer.
Simplify	This means different things in different contexts: Surds: reduce the number under the root sign to the smallest possible by removing square factors. Fractions: one fraction, cancelled down, is expected. Algebraic expressions: get rid of brackets and gather all like terms together.
Give your answer to…	This is an instruction for the accuracy of your final answer. These instructions must be followed or you will lose a mark.
Algebraically	The method you use must involve algebra e.g. you must solve an equation or simplify an algebraic expression. It is usually stated to avoid trial-and-improvement methods or reading answers from your calculator.
Justify your answer	This is a request for you to indicate clearly your reasoning. Will the examiner know how your answer was obtained?
Show all your working	Marks will be allocated for the individual steps in your working. Steps missed out may lose you marks.

In the Exam

Watch your time and pace yourself carefully. Some questions you will find harder than others. Try not to get stuck on one question as you may later run out of time. Rather return to a difficult question later. Remember also that if you have spare time towards the end of your exam, use this time to check through your solutions. Often mistakes are discovered in this checking process and can be corrected.

Become familiar with the exam instructions. The practice papers in this book have the exam instructions at the front of each exam. Also remember that there is a formuae list to consult. You will find this at the front of your exam paper. However, even though these formulae are given to you, it is important that you learn them so that they are familiar to you. If you are continuing with Mathematics next session it will be assumed that these formulae are known in next year's exam!

Read the question thoroughly before you begin to answer it – make sure you know exactly what the question is asking you to do. If the question is in sections e.g. 15a, 15b, 15c, etc. then it is often the case that answers obtained in the earlier sections are used in the later sections of that question.

When you have completed your solution read it over again. Is your reasoning clear? Will the examiner understand how you arrived at your answer? If in doubt then fill in more details.

If you change your mind or think that your solution is wrong, don't score it out unless you have another solution to replace it with. Solutions that are not correct can often gain some of the marks available. Do not miss working out. Showing step-by-step working will help you gain maximum marks even if there is a mistake in the working.

Use these resources constructively by reworking questions later that you found difficult or impossible first time round. Remember: success in a Maths exam will only come from actively trying to solve lots of questions and only consulting notes when you are stuck. Reading notes alone is not a good way to revise for your Maths exam. Always be active, always solve problems.

Good luck!

TOPIC INDEX

	A Paper 1	A Paper 2	B Paper 1	B Paper 2	C Paper 1	C Paper 2
Unit 1						
Percentages & Sig. Figs.		5		2		3, 11
Volumes of Solids	1	5		7		11
Linear Relationships		3	4		4	6
Algebraic Operations	2		6	1	1, 9	
Circles	6	2, 6	1	3		5, 12
Unit 2						
Trigonometry		9		8	5	8, 12
Simultaneous Equations		1		5		2
Graphs, Charts & Tables	7	4	2, 3		3	
Statistics	4, 7	7	2, 3	4	2	4
Applications						
Calculations in a Social Context	3	8, 10		9	7	7, 9
Logic Diagrams/Spreadsheets	5	10	5	6	6	
Formulae	8		7		8	
Further Statistics	9			10		10

In the answers, there are references to the pages of Leckie & Leckie *Intermediate 2 Maths Revision Notes* (ISBN 978-1-898890-14-0). These notes cover all of Unit 1 and Unit 2 work but do not include Applications so the references will help you learn more about any Unit 1 and Unit 2 topics you might find difficult.

Practice Exam A

Mathematics | Intermediate 2 | Units 1, 2 and Applications

Practice Papers
For SQA Exams

Exam A
Intermediate 2
Units 1, 2 and Applications
Paper 1
Non-calculator

You are allowed 45 minutes to complete this paper.

Do **not** use a calculator.

Try to answer all of the questions in the time allowed, including all of your working.

Full marks will only be awarded where your answer includes any relevant working.

FORMULAE LIST

Sine rule: $\frac{a}{\sin A} = \frac{b}{\sin B} = \frac{c}{\sin C}$

Cosine rule: $a^2 = b^2 + c^2 - 2bc \cos A$ or $\cos A = \frac{b^2 + c^2 - a^2}{2bc}$

Area of a triangle: Area = $\frac{1}{2}ab \sin C$

Volume of a sphere: Volume = $\frac{4}{3}\pi r^3$

Volume of a cone: Volume = $\frac{1}{3}\pi r^2 h$

Volume of a cylinder: Volume = $\pi r^2 h$

Standard deviation: $s = \sqrt{\frac{\Sigma(x - \bar{x})^2}{n-1}} = \sqrt{\frac{\Sigma x^2 - (\Sigma x)^2 / n}{n-1}}$, where n is the sample size.

Marks

1. The diagram shows a toy spinner.

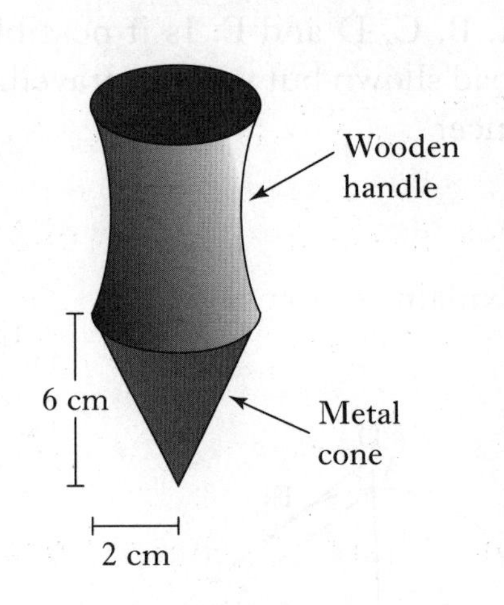

The metal cone at the base of the spinner has radius 2 centimetres and height of 6 centimetres as shown. Calculate the volume of the cone.
Take $\pi = 3{\cdot}14$ **2**

2. (*a*) Multiply out the brackets and collect like terms

$$(2x + 3y)(x - 2y)$$ **2**

(*b*) Factorise $6 + x - x^2$ **2**

3. Eilidh works for an engineering firm. She is paid a basic hourly rate of £22·60 for weekdays and at time and a half for weekends. One week she works everyday from 8 am to 2 pm. Calculate Eilidh's gross pay for that week. **3**

4. Tina and her family are playing a board game. There are two packs of cards:

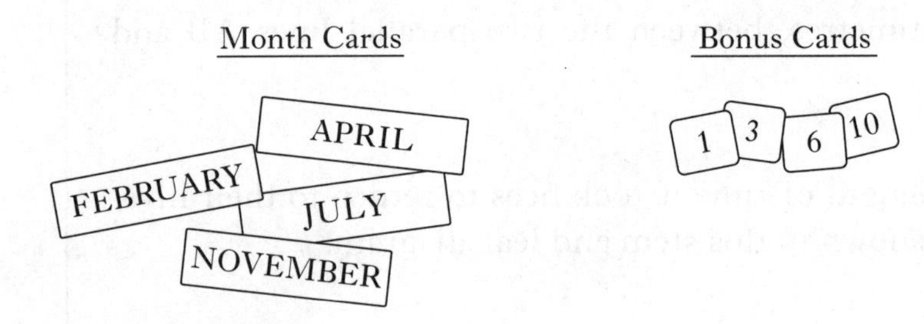

During her turn Tina picks a 'Month card' and a 'Bonus card'. This table shows all the possible outcomes for Tina:

	1	3	6	10
February	(F, 1)			
April	(A, 1)			
July	(J, 1)	(J, 3)		
November	(N, 1)			

(*a*) Copy and complete the table **1**

(*b*) What is the probability that Tina picks 'November' along with a bonus that is an odd number? **1**

Marks

5. This diagram represents the roads between the towns A, B, C, D and E. Is it possible to drive along every road shown but without travelling any road more than once?

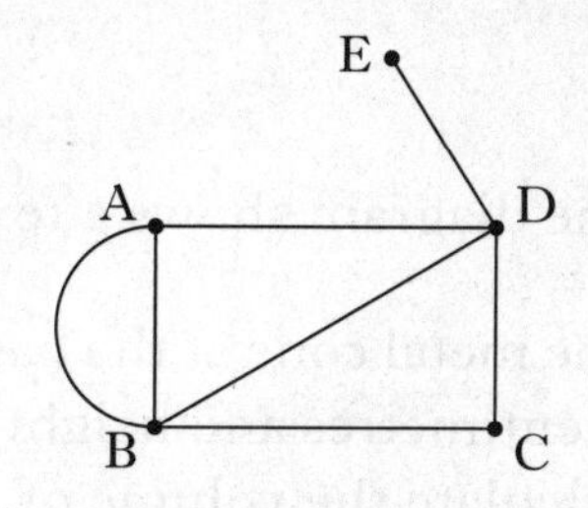

Explain your answer. **2**

6.

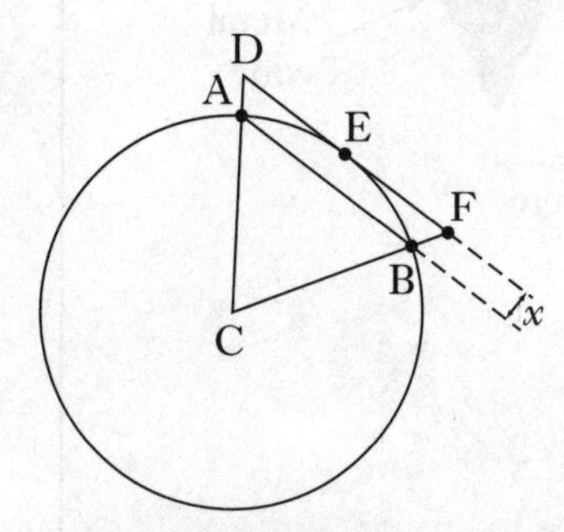

A circle, centre C, has radius of length 5 cm. Chord AB is drawn as shown and has length 6 cm.

A tangent DF is drawn to the circle with point of contact E.

Tangent DF is parallel to chord AB.

Calculate the distance, x centimetres, between the two parallel lines AB and DF. **3**

7. A Biologist was studying the length of time if took bees to return to their hive. These 'out-of-hive' times are shown in this stem and leaf diagram.

Out-of-hive times (minutes)

Stem	Leaf
0	5 6 7 7 8 9
1	2 2 5 8 9 9 9
2	3 5
3	4 7 8

$n = 18$ 2|5 represents 25 minutes

(*a*) Use the above data to calculate

(i) The median **1**

(ii) The lower quartile **1**

(iii) The upper quartile **1**

(iv) The semi-interquartile range **1**

(*b*) For a second hive the semi-interquartile range was found to be 3 minutes. Make a valid comparison concerning the 'out-of-hive' times for the two hives. Give a reason for your answer. **1**

Marks

8. The surface area, A square centimetres, of a regular pentagonal prism is given by the formula.

$A = 5as + 5sh$

where s centimetres is the length of the side of the pentagon
h centimetres is the height of the prism
a centimetres is the distance from the centre of the pentagon to the sides.

(*a*) Calculate A when $a = 2$, $s = 3{\cdot}8$ and $h = 7$ **2**

(*b*) Calculate s when $A = 270$, $a = 3$ and $h = 6$ **3**

9. One hundred pupils at a secondary school were chosen to sit a numeracy test.

The results of the test are shown in this table.

Mark	Frequency	Cumulative frequency
1–10	7	
11–20	12	
21–30	19	
31–40	24	
41–50	20	
51–60	9	
61–70	7	
71–80	2	

(*a*) Copy and complete the table. **1**

(*b*) Using this data, draw a cumulative frequency curve on squared paper. **2**

(*c*) Use the curve you have drawn to estimate the median mark scored in the test. **1**

[End of Question Paper]

Marks

8. The surface area, A square centimetres, of a regular pentagonal prism is given by the formula

$A = 5as + 5sh$

where s centimetres is the length of the side of the pentagon
h centimetres is the height of the prism
a centimetres is the distance from the centre of the pentagon to the sides.

(a) Calculate A when $a = 2$, $s = 3.8$ and $h = 7$. **2**

(b) Calculate s when $A = 270$, $a = 4$ and $h = 6$. **3**

9. One hundred pupils at a secondary school were chosen to sit a numeracy test.

The results of the test are shown in this table.

Mark	Frequency	Cumulative frequency
1–10	7	
11–20	12	
21–30	19	
31–40	[illegible]	
41–50	20	
51–60	8	
61–70	7	
[illegible]	[illegible]	

(a) Copy and complete the table. **1**

(b) Using this data, draw a cumulative frequency curve on squared paper. **2**

(c) Use the curve you have drawn to estimate the median mark scored in the test. **1**

[End of Question Paper]

Mathematics | Intermediate 2 | Units 1, 2 and Applications

Practice Papers
For SQA Exams

Exam A
Intermediate 2
Units 1, 2 and Applications
Paper 2

You are allowed 1 hour, 30 minutes to complete this paper.

A calculator can be used.

Try to answer all of the questions in the time allowed, including all of your working.

Full marks will only be awarded where your answer includes any relevant working.

FORMULAE LIST

Sine rule: $\frac{a}{\sin A} = \frac{b}{\sin B} = \frac{c}{\sin C}$

Cosine rule: $a^2 = b^2 + c^2 - 2bc \cos A$ or $\cos A = \frac{b^2 + c^2 - a^2}{2bc}$

Area of a triangle: Area = $\frac{1}{2} ab \sin C$

Volume of a sphere: Volume = $\frac{4}{3}\pi r^3$

Volume of a cone: Volume = $\frac{1}{3}\pi r^2 h$

Volume of a cylinder: Volume = $\pi r^2 h$

Standard deviation: $s = \sqrt{\frac{\Sigma(x - \bar{x})^2}{n-1}} = \sqrt{\frac{\Sigma x^2 - (\Sigma x)^2 / n}{n-1}}$, where n is the sample size.

Marks

1. An hotel has 22 rooms.

The rooms are of two types: single or double.

A single room costs £80 per night and a double room costs £120 per night

Let x be the number of single rooms and let y be the number of double rooms

(*a*) Write down an equation in x and y which satisfies the information given above. **1**

(*b*) When the hotel is full the takings for that night are £2320. Write down a second equation in x and y which satisfies this condition. **1**

(*c*) How many single rooms and how many double rooms does the hotel have? **4**

2. PQ is a tangent to the circle with centre O and touches the circle at point A.

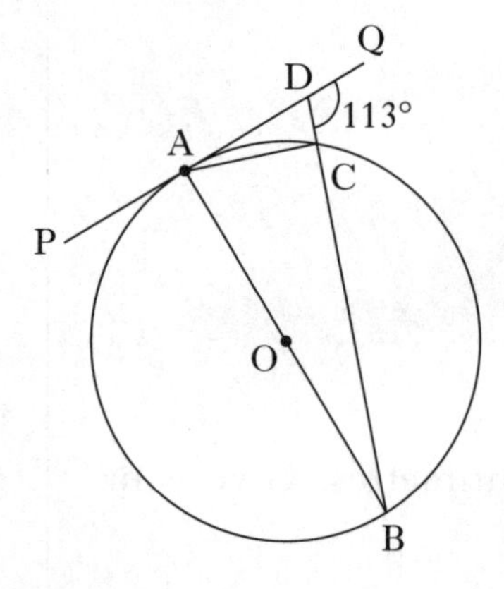

AB is a diameter of the circle.

Angle BDQ = 113° as shown.

Find the size of angle BAC. **3**

3. Find the equation of the line shown in the diagram which passes through the points (–2, 4) and (0, 3). **3**

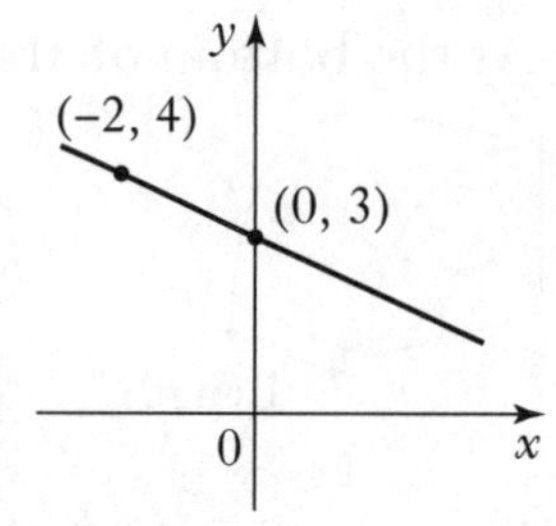

4. A vacuum cleaner production company was developing a new cleaner.

They had four prototypes: type A, type B, type C and type D.

A sample of households were asked to try all four cleaners for a month and then state their preference at the end of the trial.

Their preferences are shown in this table:

Type	Frequency
A	27
B	9
C	27
D	18

Construct a pie chart to illustrate this information. Show all your working. **3**

Marks

5.

35 cm

An antique television is housed in a wooden cabinet in the shape of a prism 35 cm deep as shown in the diagram above.

The front of the cabinet is rectangular with a semicircle added at the top. The following diagram shows the measurments:

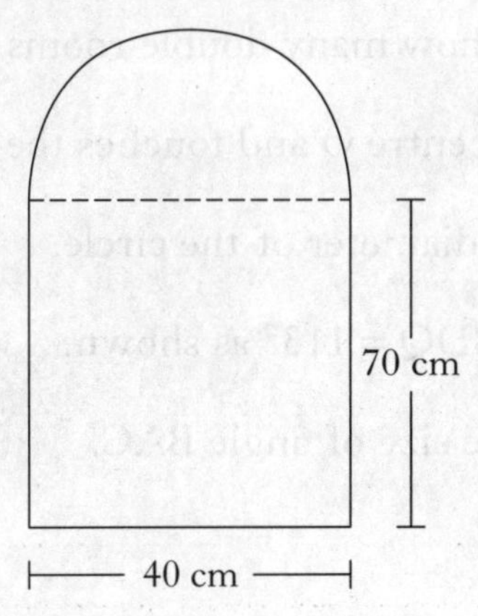

(*a*) Find the volume of the wooden cabinet in cubic centimetres. Give your answer correct to two significant figures. **4**

At the bottom of the cabinet there are two drawers.

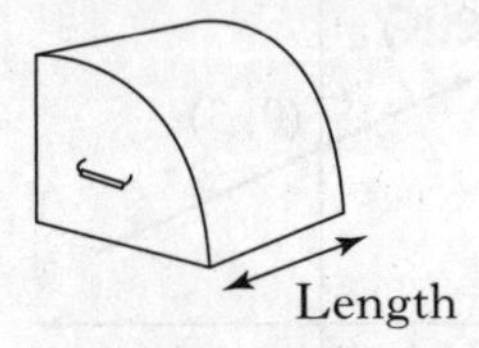

The uniform cross-section of each drawer is a quarter of a circle.

The volume of each drawer is 10 000 cm^3.

(*b*) Find the depth of the drawer. **3**

6. This clock face is showing the time four o'clock.

The minute hand, the longer of the two hands, is 8 centimetres in length.

The clock face is shown in this diagram 35 minutes later at 25 minutes to five o'clock.

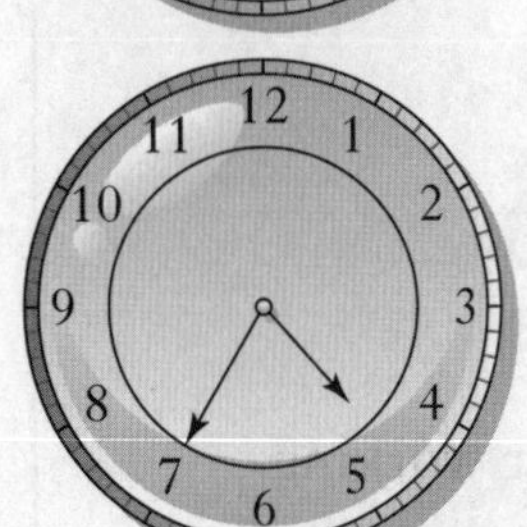

Find the distance the tip of the minute hand travelled in these 35 minutes. **4**

Marks

7. A Quality Control Inspector selects a random sample of seven matchboxes produced by Machine A and records the number of matches in each box:

54 45 51 50 48 53 49

(*a*) For the given data calculate:

(i) the mean — 1

(ii) the standard deviation — 3

Show clearly all your working.

(*b*) Machine B was also sampled. The data gave a mean of 52 matches and a standard deviation of 1·6 matches. Compare the results for the two machines justifying your comparisons. — 2

8. Olivia is a sales assistant and is paid a basic wage of £500 per week. In addition she is paid a commission of 2·5% on her sales for the week.

(*a*) On a week where her sales were £2400 what will her total earnings for the week be? — 2

A new commission scheme is introduced

Sales	Rate of Commission
Less than £2500	2·5%
£2500 to £5000	2·75%
More than £5000	3·0%

(*b*) On the first week on this new scheme Olivia earned a total of £710. What were her sales during this first week? — 4

9. Mia is constructing a kite from fibre-glass rods and plastic sheeting. The frame ABCD has measurements as shown:

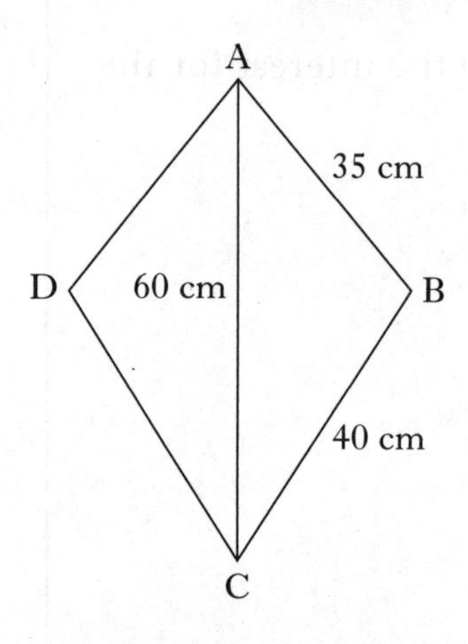

(*a*) Calculate the size of angle ACB.
Do not use a scale drawing. — 3

(*b*) Calculate the area of plastic sheeting she needs to cover the frame. — 3

Marks

10. Aimee plans to take out a loan for £4500, she creates a spreadsheet to compare the cost of a loan from two different banks.

Bank A calculates interest monthly (0·6% per month)

Bank B calculates interest quarterly (1·9% per 3 months)

	A	B	C	D	E	F
1	Bank A			Bank B		
2						
3	Interest per month	0·6	%	Interest per 3 months	1·9	%
4						
5	Amount owed	£4500·00		Amount owed	£4500·00	
6	After 1 month	£4527·00				
7	After 2 months	£4554·16				
8	After 3 months	£4581·49		After 3 months	£4585·50	
9	After 4 months	£4608·98				
10	After 5 months	£4636·63				
11	After 6 months	£4664·45		After 6 months	£4672·62	
12	After 7 months	£4692·44				
13	After 8 months	£4720·59				
14	After 9 months	£4748·91		After 9 months	£4761·40	
15	After 10 months	£4777·41				
16	After 11 months	£4806·07				
17	After 12 months			After 12 months		
18						
19	Interest for 1 year:			Interest for 1 year:		

(*a*) Write down suitable formulae to enter in B17 and E17 to calculate the amount owed to each bank after 12 months. **2**

(*b*) The formula = B17 – B5 is entered in cell B19. What result will appear in B19? **1**

(*c*) Write down the formula to enter in cell E19 to calculate the interest for the year's loan from Bank B. **1**

(*d*) Which loan is more expensive and by how much? **2**

[End of Question Paper]

Practice Exam B

Mathematics | Intermediate 2 | Units 1, 2 and Applications

Practice Papers
For SQA Exams

Exam B
Intermediate 2
Units 1, 2 and Applications
Paper 1
Non-calculator

You are allowed 45 minutes to complete this paper.

Do **not** use a calculator.

Try to answer all of the questions in the time allowed, including all of your working.

Full marks will only be awarded where your answer includes any relevant working.

FORMULAE LIST

Sine rule: $\dfrac{a}{\sin A} = \dfrac{b}{\sin B} = \dfrac{c}{\sin C}$

Cosine rule: $a^2 = b^2 + c^2 - 2bc \cos A$ or $\cos A = \dfrac{b^2 + c^2 - a^2}{2bc}$

Area of a triangle: Area = $\frac{1}{2}ab \sin C$

Volume of a sphere: Volume = $\frac{4}{3}\pi r^3$

Volume of a cone: Volume = $\frac{1}{3}\pi r^2 h$

Volume of a cylinder: Volume = $\pi r^2 h$

Standard deviation: $s = \sqrt{\dfrac{\Sigma(x - \bar{x})^2}{n - 1}} = \sqrt{\dfrac{\Sigma x^2 - (\Sigma x)^2 / n}{n - 1}}$, where n is the sample size.

		Marks

1.

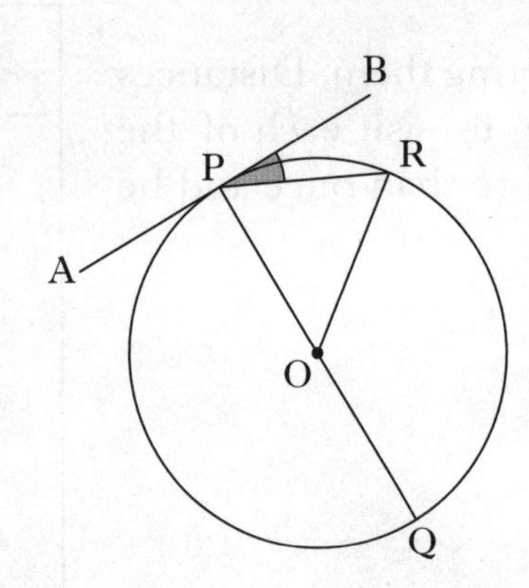

AB is a tangent to the circle, centre O, with point of contact P.

POQ is a diameter and PR is a chord.

The shaded angle BPR = 22°

Calculate the size of angle ROQ. **3**

2. The shoe sizes of a group of 20 students were recorded.

6	7	8	7½	9	7½	7	8	7½	9
7½	7½	8	7	6½	8	7½	6½	8½	7½

(*a*) Construct a frequency table for this data and add a cumulative frequency column. **2**

(*b*) What is the probability that one of the students picked at random has shoe size greater than 7½? **1**

3. The following data gives the ages of the workers on oil rig Alpha:

21	22	24	23	27	23	23
24	21	22	23	23	21	29
23	19	23	21	24	22	23

(*a*) Construct a dotplot for this data. **2**

(*b*) For this data find:

(i) The median; **1**

(ii) The lower quartile; **1**

(iii) The upper quartile. **1**

(*c*) For the workers on oil rig Beta the semi-interquartile range of their ages is 3·5 years. Make an appropriate comment on the distribution of the ages of the workers on the two oil rigs. **2**

4.

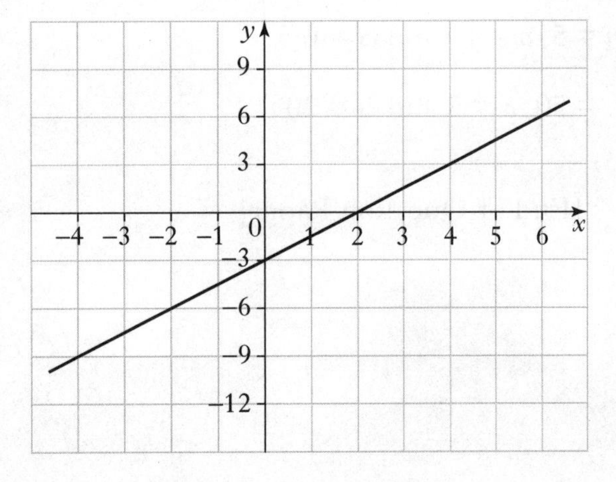

Find the equation of the line shown in the diagram. **3**

Marks

5. The diagram represents 5 towns in Fife and the roads connecting them. Distances are given in kilometres. A cyclist in Austruther is planning to visit each of the other 4 towns. His route must not go through any town more than once and he does not need to return to Austruther.

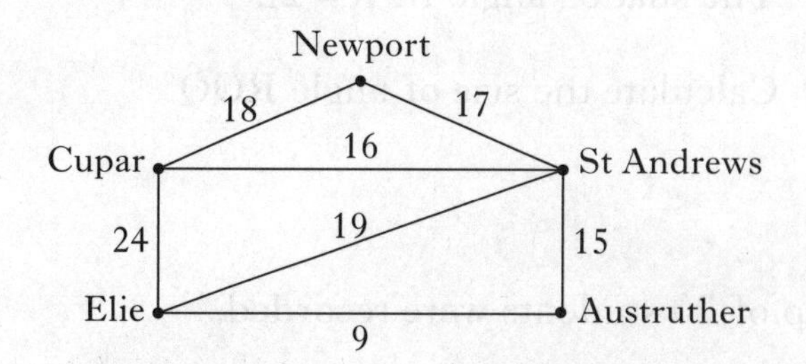

(*a*) Draw a tree diagram to show all his possible route **3**

(*b*) Which is the shortest route? **3**

Show clearly all your working.

6. 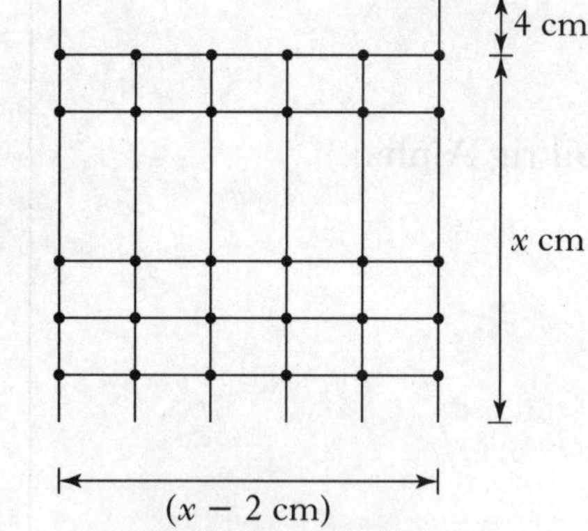

The diagram shows the measurements of a metal gate which is made up of 6 vertical rods and 5 horizontal rods.

(*a*) Show that the total length of rod, L cm, used to make the gate is given by $L = 11x - 2$ **2**

(*b*) Find x if the total length of rod in the gate is 713 cm. **1**

7. The length, L, of a curve called an Astroid is given by the formula:

$$L = \frac{Sa(n-1)}{n}$$

(*a*) Calculate L when $n = 5$, $a = 10$ and $S = 0{\cdot}5$ **2**

(*b*) Calculate n when $L = 50$, $a = 3$ and $S = 20$ **3**

[End of Question Paper]

Mathematics | Intermediate 2 | Units 1, 2 and Applications

Practice Papers
For SQA Exams

Exam B
Intermediate 2
Units 1, 2 and Applications
Paper 2

You are allowed 1 hour, 30 minutes to complete this paper.

A calculator can be used.

Try to answer all of the questions in the time allowed, including all of your working.

Full marks will only be awarded where your answer includes any relevant working.

FORMULAE LIST

Sine rule: $\dfrac{a}{\sin A} = \dfrac{b}{\sin B} = \dfrac{c}{\sin C}$

Cosine rule: $a^2 = b^2 + c^2 - 2bc \cos A$ or $\cos A = \dfrac{b^2 + c^2 - a^2}{2bc}$

Area of a triangle: Area $= \frac{1}{2}ab \sin C$

Volume of a sphere: Volume $= \frac{4}{3}\pi r^3$

Volume of a cone: Volume $= \frac{1}{3}\pi r^2 h$

Volume of a cylinder: Volume $= \pi r^2 h$

Standard deviation: $s = \sqrt{\dfrac{\Sigma(x-\bar{x})^2}{n-1}} = \sqrt{\dfrac{\Sigma x^2 - (\Sigma x)^2/n}{n-1}}$, where n is the sample size.

Marks

1. (*a*) Factorise

$2x^2 + x - 28$ **2**

(*b*) Multiply out the brackets and collect like terms

$(2a - 3b)(3a + 2b) + 2ab$ **3**

2. During a flu epidemic 6400 cases were recorded on Monday.

The number of cases was expected to rise by 28·5% each day.

How many cases are expected by Thursday in the same week?

Give your answer correct to three significant figures. **3**

3.

Three sheep pens are constructed from fencing. Each pen is an equal sector of a semicircle with diameter 40 metres.

Calculate the total length of fencing required to enclose pen 1, the shaded pen in the diagram. **3**

4. Temperature readings were taken each day at noon. The temperatures in °C are:

8 12 13 12 11 11 10

(*a*) Use an appropriate formula to calculate the mean and standard deviation of these temperatures. **4**

Show clearly all your working

(*b*) It was discovered that the digital thermometer used to collect the data had a malfunction and consistently recorded temperatures 2°C higher than they actually were. State the mean and standard deviation of the actual temperatures. **2**

Marks

5. Hector and Angus run a Traditional Scottish Fiddle Music Centre. On Tuesdays and Thursdays they offer tuition classes at two levels of difficulty: 'Beginners' and 'Advanced'

(*a*) On Tuesday 3 people turned up for the 'Beginners' class and 5 people turned up for the 'Advanced' class. The tuition fees collected that evening totalled £190. Let £x be the fee for the 'Beginners' class and let £y be the fee for the 'Advanced class'.

Write down an equation in x and y which satisfies the above condition. **1**

(*b*) On Thursday the turn out was: 4 people for the 'Beginners' class and 2 people for the 'Advanced' class. Total takings for that evening were £132.

Write down a second equation in x and y which satisfies this condition. **1**

(*c*) Calculate the fee for the 'Beginners' class and the fee for the 'Advanced' class. **4**

6. A carhire firm calculates its hire charge from:

- The number of days hired
- The number of miles travelled

They use the following flowchart:

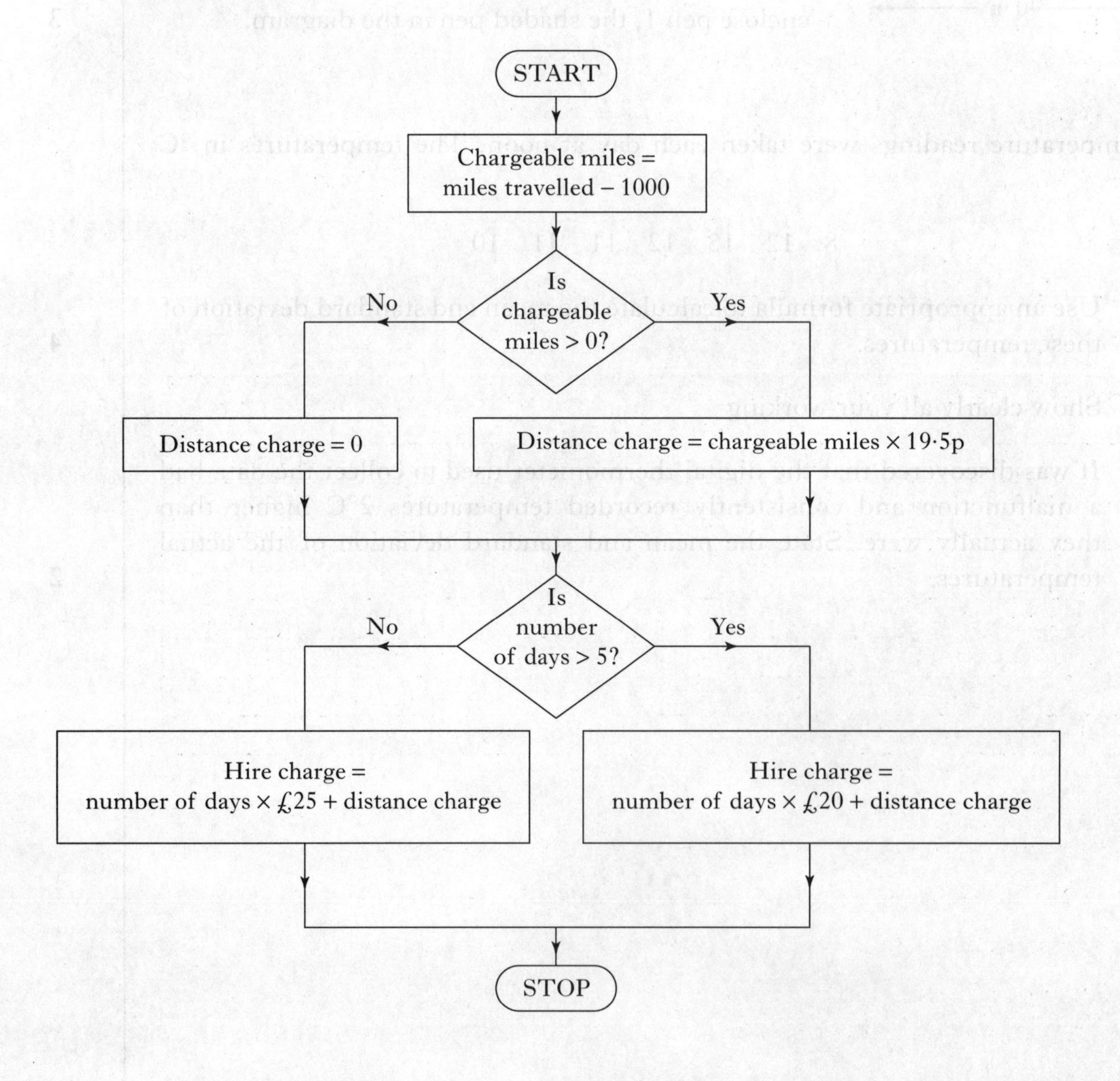

Marks

Jamie • hires a car for 5 days
• travels a distance of 2500 miles

Calculate his hire charge. **3**

7. The local 'Thai Cuisine' restaurant uses two types of frying pan: the traditional 'wok' in the shape of a hemisphere and a normal cylindrical pan. The measurements are shown in this diagram:

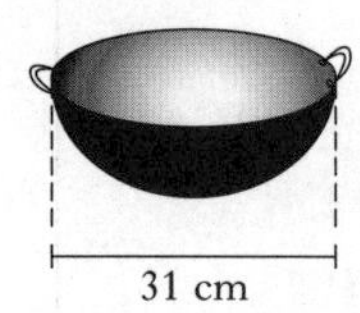

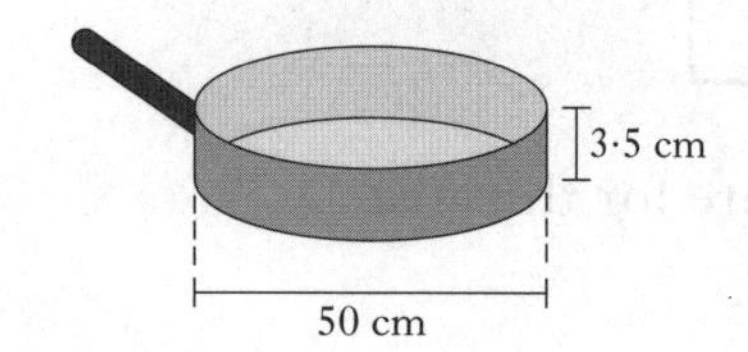

Which container has the larger capacity? **5**

8. The diagram shows a metal plate (shaded in the diagram) in the shape of a quadrilateral ABCD with measurements as shown.

Calculate:

(*a*) The distance between the two vertices B and D of the metal plate (the broken line in the diagram). Do not use a scale drawing. **3**

(*b*) The area of the plate (the shaded area in the diagram). **4**

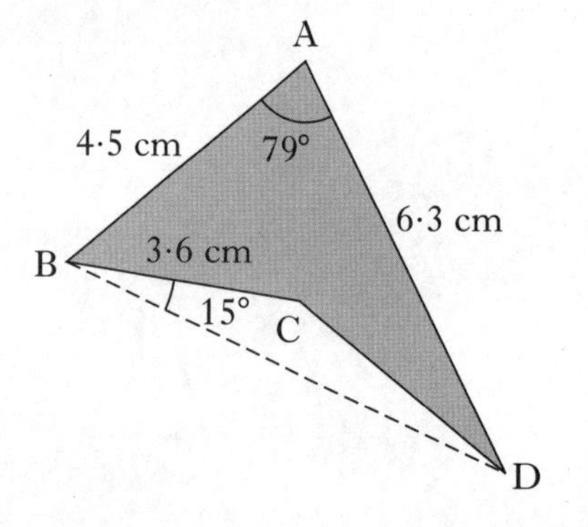

9. Lighter Loans specialises in long term loans. The table below shows the monthly repayments to be made with or without payment protection insurance.

Monthly loan Repayments over 5 to 15 years with and without Payment Protection Insurance							
		60 months		**120 months**		**180 months**	
Amount of loan	**APR %**	**without insurance**	**with insurance**	**without insurance**	**with insurance**	**without insurance**	**with insurance**
£5000	12·9	£111·80	£144·22	£72·40	£89·38	£60·75	£72·70
£10000	12·9	£223·60	£288·44	£144·80	£178·76	£121·50	£145·40
£15000	11·4	£325·26	£418·10	£205·64	£251·31	£169·40	£200·39
£20000	11·4	£433·70	£559·47	£274·18	£335·06	£225·86	£267·18
£25000	11·4	£542·13	£699·34	£342·73	£418·85	£282·33	£333·98

Jack takes out two loans, one for £5000 over 5 years and one for £15000 over 10 years. Both loans are without payment protection insurance.

(*a*) State his monthly repayment to the company for the £5000 loan. **1**

(*b*) Calculate the total cost to Jack of the £5000 loan over the 5 years. **1**

(*c*) Would it cost Jack less or more to take out of £20000 loan, without payment protection, over 10 years, than to take out these two separate loans? **2**

Marks

10. The scores for a class test are shown in this frequency table:

Score	Frequency
0–6	1
7–13	4
14–20	11
21–27	8
28–34	3

Calculate the mean score for this test. **5**

[End of Question Paper]

Practice Exam C

Mathematics | Intermediate 2 | Units 1, 2 and Applications

Practice Papers
For SQA Exams

Exam C
Intermediate 2
Units 1, 2 and Applications
Paper 1
Non-calculator

You are allowed 45 minutes to complete this paper.

Do **not** use a calculator.

Try to answer all of the questions in the time allowed, including all of your working.

Full marks will only be awarded where your answer includes any relevant working.

FORMULAE LIST

Sine rule: $\dfrac{a}{\sin A} = \dfrac{b}{\sin B} = \dfrac{c}{\sin C}$

Cosine rule: $a^2 = b^2 + c^2 - 2bc \cos A$ or $\cos A = \dfrac{b^2 + c^2 - a^2}{2bc}$

Area of a triangle: Area = $\frac{1}{2} ab \sin C$

Volume of a sphere: Volume = $\frac{4}{3}\pi r^3$

Volume of a cone: Volume = $\frac{1}{3}\pi r^2 h$

Volume of a cylinder: Volume = $\pi r^2 h$

Standard deviation: $s = \sqrt{\dfrac{\Sigma(x - \bar{x})^2}{n-1}} = \sqrt{\dfrac{\Sigma x^2 - (\Sigma x)^2 / n}{n-1}}$, where n is the sample size.

Marks

1. (*a*) Multiply out the brackets and collect like terms

$3k + (2k - 3)(k - 4)$ **3**

(*b*) Factorise

$7m^2 + 54m - 16$ **2**

2. All the passengers on a bus were asked how long they had waited before they got on the bus. The stem and leaf diagram below shows the results:

Stem	Leaf
0	2 2 3 5 6 6 8 9
1	0 0 1 2 8
2	1 6
3	0

$n = 16$ 2|6 represents 26 minutes

What is the probability that a passenger on the bus, chosen at random, waited more than 15 minutes? **1**

3. This stem and leaf diagram shows the time taken, in minutes, by a group of hill walkers to complete a mountain walk.

Stem	Leaf
8	5
9	3
10	1 2 7 9
11	0 3 4 7 9 9
12	0

$n = 13$ 10|7 represents 107 minutes

(*a*) For the given data calculate:

(i) The median **1**

(ii) The lower and upper quartiles **2**

(*b*) Draw a boxplot to illustrate this data. **2**

4. (*a*) State the gradient of the line shown in the diagram. **1**

(*b*) Find the equation of the line. **2**

(*c*) Find the coordinates of the point where the line $y = x - 1$ meets this line. **2**

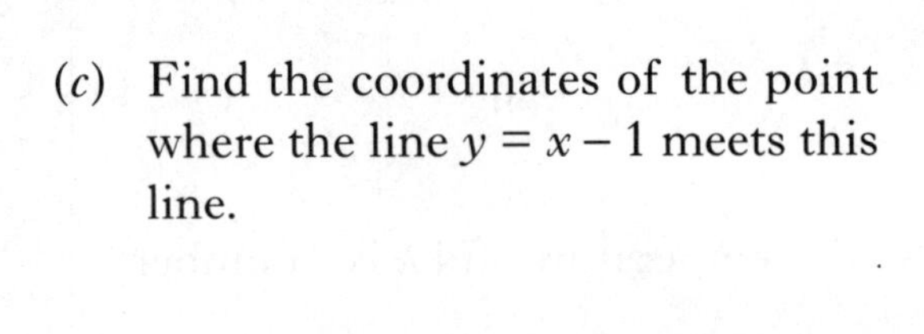

Marks

5. Given that cos 18° = 0·951 to 3 decimal places state a value for x other than 18 for which

$\cos x° = 0{\cdot}951,\ 0 \leq x \leq 360$ **1**

6. A network is Hamiltonian if you can draw a route along the edges of the network that visits every node exactly once without lifting your pencil and return to your starting node.

For example the network shown on the right is Hamiltonian.

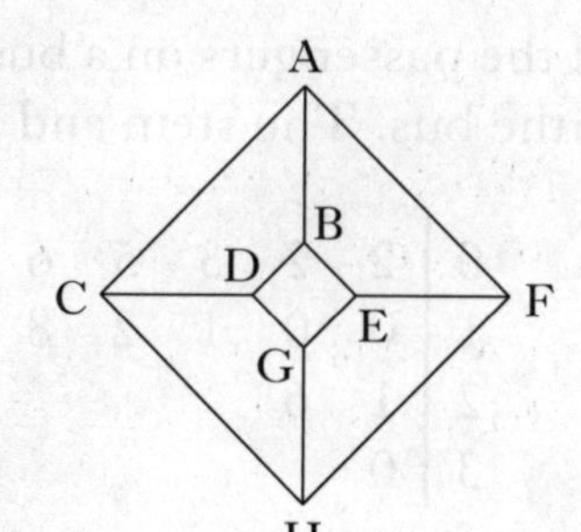

A route that shows this is:

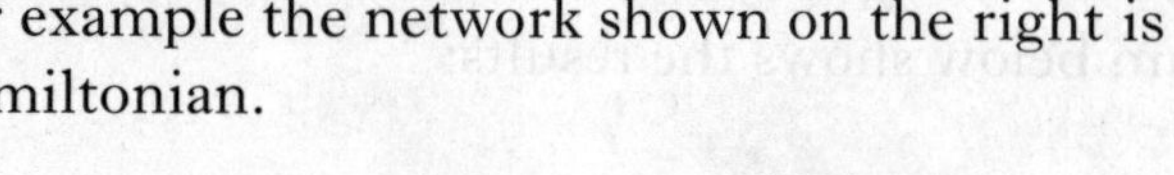

$A \to B \to D \to G \to E \to F \to H \to C \to A$

Is the following network Hamiltonian?

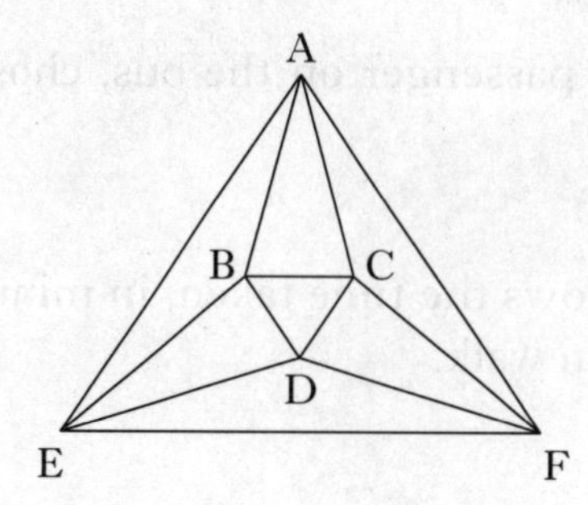

Explain your answer **2**

7. This spreadsheet calculates the wages of employees in a company:

	A	B	C	D	E	F
1	Employee	Basic Rate	Hours	Pay	Hours (double time)	Overtime pay
2	Nick Evans	£8·40	32	£260·80	5	£94·00
3	Neil Brand	£6·25	35	£218·75	3	£37·50

(*a*) The result of the formula = D2 + F2 is entered in cell G2. What will appear in cell G2? **1**

(*b*) Write down the formula that is entered in cell F3 to calculate Neil Brand's overtime pay. **1**

8. A metal rod expands when it is heated. Its new Length, L cm, is given by the formula:

$$L = l(1 + kt)$$

where l cm is the original length, t°c is the rise in temperature and k is a number that is determined by the type of metal.

(*a*) Calculate L when $l = 12$, $k = 0{\cdot}004$ and $t = 5$ **3**

(*b*) Calculate t when $l = 5$, $k = 0{\cdot}007$ and $L = 5{\cdot}28$ **3**

	Marks

9. A drawing with dimensions

$(x + 1)$ centimetres $\times$ $(x - 2)$ centimetres

is surrounded by a frame with dimensions

$(x + 3)$ centimetres $\times$ x centimetres

The drawing and frame are shown in this diagram:

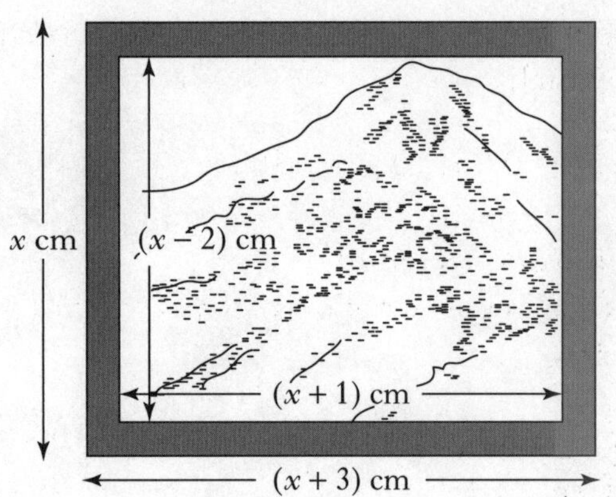

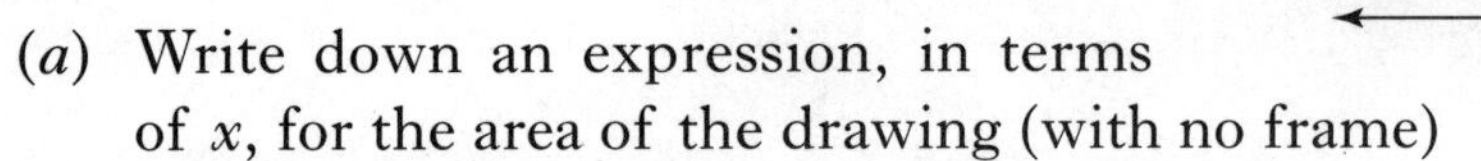

(*a*) Write down an expression, in terms of x, for the area of the drawing (with no frame) **1**

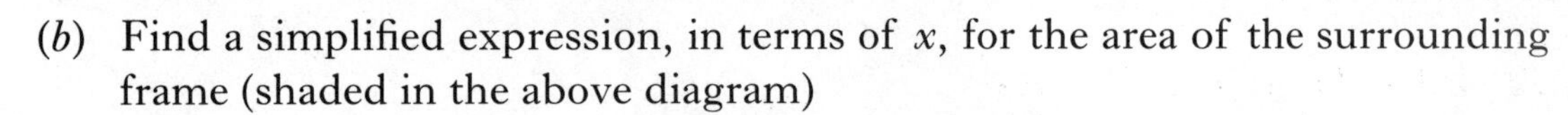

(*b*) Find a simplified expression, in terms of x, for the area of the surrounding frame (shaded in the above diagram) **2**

[End of Question Paper]

Mathematics | Intermediate 2 | Units 1, 2 and Applications

Practice Papers
For SQA Exams

Exam C
Intermediate 2
Units 1, 2 and Applications
Paper 2

You are allowed 1 hour, 30 minutes to complete this paper.

A calculator can be used.

Try to answer all of the questions in the time allowed, including all of your working.

Full marks will only be awarded where your answer includes any relevant working.

FORMULAE LIST

Sine rule: $\dfrac{a}{\sin \mathrm{A}} = \dfrac{b}{\sin \mathrm{B}} = \dfrac{c}{\sin \mathrm{C}}$

Cosine rule: $a^2 = b^2 + c^2 - 2bc \cos \mathrm{A}$ or $\cos \mathrm{A} = \dfrac{b^2 + c^2 - a^2}{2bc}$

Area of a triangle: Area = $\frac{1}{2}ab \sin \mathrm{C}$

Volume of a sphere: Volume = $\frac{4}{3}\pi r^3$

Volume of a cone: Volume = $\frac{1}{3}\pi r^2 h$

Volume of a cylinder: Volume = $\pi r^2 h$

Standard deviation: $s = \sqrt{\dfrac{\Sigma(x - \bar{x})^2}{n - 1}} = \sqrt{\dfrac{\Sigma x^2 - (\Sigma x)^2 / n}{n - 1}}$, where n is the sample size.

Marks

1. Thomas is selling his house by auction. The cost of the auction sale is made up of three parts:

- A catalogue entry fee of £325
- A commission charge of 1·2% of the selling price of the property
- VAT at 17·5%

His house was successfully auctioned. If the selling price was £180 500 calculate the cost of the auction sale before VAT was added. **2**

2. A weightlifter owns two types of circular weights:

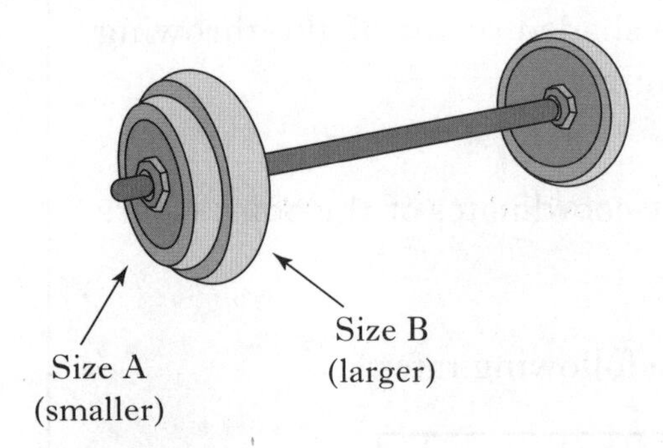

Here are two arrangements of Size A and Size B weights and their corresponding total weight:

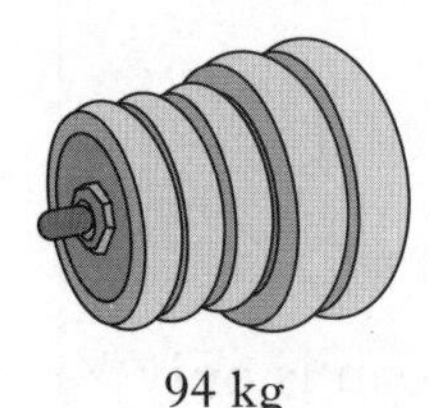

94 kg

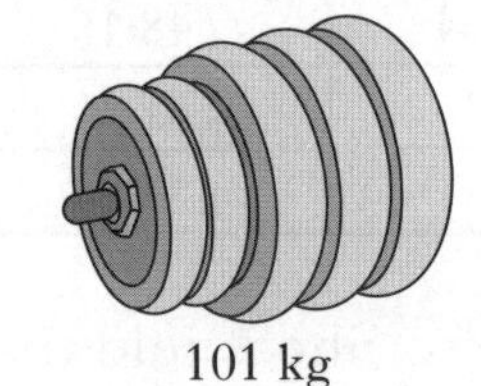

101 kg

Find the weight of one size A weight and the weight of one size B weight. **6**

3. The number of bacteria in a laboratory culture dropped by 22% per minute after being treated with an antibiotic. If there where 250000 bacteria in the culture at 10 am after treatment, how many were left by 10·03 am? **3**

4. A ream of paper contains 500 sheets.

A sample of five reams was checked and the number of sheets recorded:

503 504 497 495 506

For this data the mean number of sheets is 501.

(*a*) Calculate the standard deviation **3**

Show clearly all your working

(*b*) It was felt by the producer that there was too much variation in the number of sheets in the sample.

Marks

The machine that bundled the reams was adjusted and a new sample was tested. The new mean and standard deviation for this sample were 502 sheets and 3·5 sheets respectively. Did the adjustment produce less variation?

Give a reason for your answer **1**

5. 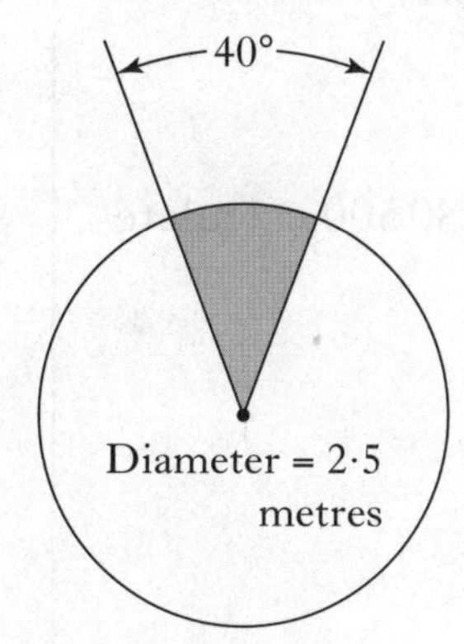

The diagram shows the plan of a Discus throwing circle for an athletics field.

The circle has diameter of 2·5 metres

For a valid throw the discus must land within the extended lines of the 40° sector of the circle.

Calculate the area of the shaded sector of the throwing circle as shown in the diagram. **3**

6. A straight line has equation $2x - 3y = 12$. Find the coordinates of the point where it crosses the y-axis **2**

7. The loan For Living company offers loans at the following rates:

MONTHLY REPAYMENTS ON A LOAN OF £5000				
APR	**60 months**	**120 months**	**180 months**	**240 months**
9·4%	£103·93	£63·50	£50·89	£45·18
10·4%	£106·16	£66·00	£53·64	£48·16
11·4%	£108·43	£68·54	£56·46	£51·21
12·4%	£110·67	£71·11	£59·31	£54·30

Cathy Hall borrows £17500 from this company. She is told that the loan is only available at an annual percentage rate (APR) of 10·4% and for 180 months.

Use the table to calculate the total cost of the loan. **4**

8. The Ecology team at a University are dividing an area of terrain into triangular sections to study the eco-systems.

Their base camp is at Altair (A on the diagram). They use three rocks (B, C and D on the diagram) to mark out two of the triangular sections.

Bearings and distances are:
Rock B is 12 km from Base camp A on a bearing of 105°

Rock C is 10 km from rock B on a bearing of 205°

(*a*) Calculate the size of angle ABC **1**

(*b*) Calculate the distance of rock C from the base camp at A. **3**

Angle DAC is 40° with rock D being 8 km from the Base Camp at A

(*c*) Calculate the area of the triangular section ACD **2**

Marks

9. Here is part of Jennifer Smart's Credit Card Statement showing her transactions for the first month of her new account:

CREDIT CARD STATEMENT		AUGUST	STATEMENT 1
PURCHASES	**£**	**DATE**	**NOTES**
Esse Stores H & M Sport KK Mic Lal Ferries	23·52 112·00 57·13 126·05	2 AUG 2009 12 AUG 2009 12 AUG 2009 21 AUG 2009	• Minimum monthly payment is 3% of the total due • Monthly interest is 2·2% of total due (APR 29·8%) after monthly payment is made
Total DUE:	318·70		

(*a*) Jennifer pays the minimum monthly payment. How much does she pay? **2**

(*b*) In September her purchases amount to £150. Calculate the total due that will appear on her September statement. **3**

10. The golf scores of players on the Canmore Golf Course one Tuesday were recorded.

Score	Frequency
68–70	1
71–73	3
74–76	8
77–79	17
80–82	12
83–85	9
86–88	7
89–91	5
92–94	1
95–97	1

(*a*) Construct a cumulative frequency column for this data. **1**

(*b*) Using squared paper, draw a cumulative frequency diagram for this data. **3**

(*c*) Use your diagram to estimate the median score of all the players on this day. **1**

11. The diagram on the right shows a plumb-line used by bricklayers to ensure that their constructions are vertical.

The weight at the bottom is in the shape of a cone and is made of metal.

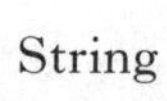

Here are the dimensions of the cone:

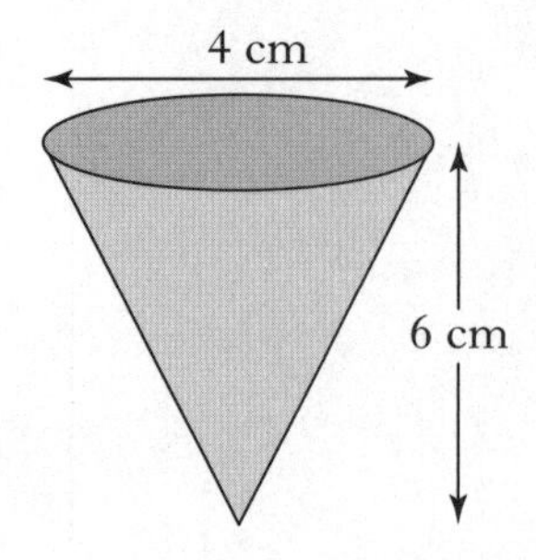

(*a*) Calculate the volume of this conical weight.

Give your answer correct to three significant figures. **3**

Marks

(*b*) The weight is redesigned into the shape of half of a cylinder as shown in the diagram on the right. The same metal is used for this new weight and has the same volume as the old conical weight.

4 cm

height

Calculate the height of this new weight. **3**

12. The diagram shows a pentagon inscribed in a circle, centre O, with radius 5 cm. All sides of the pentagon are equal in length.

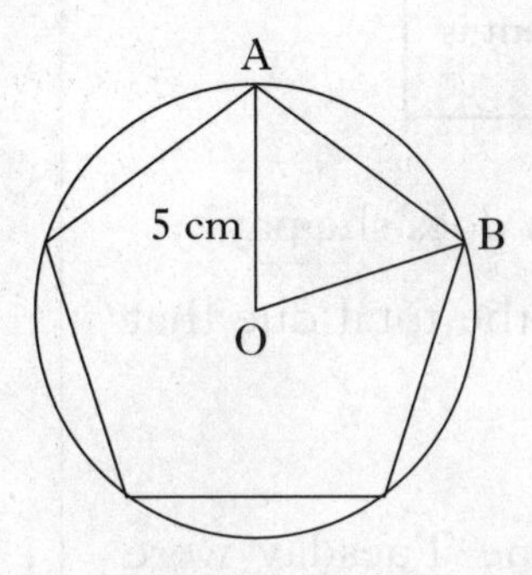

(*a*) Find the size of angle AOB **1**

(*b*) Calculate the length of AB, one of the sides of the pentagon. **3**

[End of Question Paper]

Worked Answers

Q 1.
Volume = $\frac{1}{3}\pi r^2 h$
with $\pi = 3{\cdot}14$, $r = 2$ and $h = 6$
so Volume = $\frac{1}{3} \times 3{\cdot}14 \times 2^2 \times 6$ ✓
$= \frac{1}{3} \times 6 \times 2^2 \times 3{\cdot}14$
$= 2 \times 4 \times 3{\cdot}14$
$= 8 \times 3{\cdot}14$
$= 25{\cdot}12$ cm^3 ✓

2 marks

Substitution
- The formula $V = \frac{1}{3}\pi r^2 h$ is given to you on the formulae sheet during your exam
- The first mark is given for the correct substitution of the numbers into the formula
- You do not have access to a calculator in Paper 1 so all calculations must be done on your answer paper. This is why the value $\pi = 3{\cdot}14$ is given for this calculation. You must not use any other value if you are asked to use this value.

Calculation
- The second mark is for correctly calculating the volume
- No marks are awarded for rounding in this question. This is generally the case in a question that does not specify the accuracy required for the answer.

See Int 2 Notes, Section 2, p. 8

Q 2. *(a)*
$(2x + 3y)(x - 2y)$
$= 2x^2 - 4xy + 3xy - 6y^2$ ✓
$= 2x^2 - xy - 6y^2$ ✓

2 marks

Starting
- Any two correct terms among $2x^2$, $-4xy$, $3xy$ and $-6y^2$ will gain you this first mark
- The pattern used is:

F L
(a + b) (c + d)
I
O

remember FOIL

F: Firsts O: Outsides I: Insides L: Lasts

$2x \times x$	$2x \times (-2y)$	$3y \times x$	$3y \times (-2y)$
$2x^2$	$-4xy$	$3xy$	$-6y^2$

Completing
- Gather 'like terms': $-4xy + 3xy = -xy$

See Int 2 Notes, Section 4, p. 13

Q 2. *(b)*
$6 + x - x^2$ ✓
$= (2 + x)(3 - x)$ ✓

2 marks

1st factor
- There are two choices for the 'FIRSTS': $2 \times 3 = 6$ or $1 \times 6 = 6$. For the 'LASTS' there is only one choice: $x \times x = x^2$
- You should 'multiply out' your answer using 'FOIL' to check that you get $6 + x - x^2$— if not then you're wrong!

2nd factor
- If you mix up the signs: $(2 - x)(3 + x)$ you will still gain 1 mark.

Q 3.
From 8am to 2pm is 6 hours.
Basic Rate Pay:
5 days of 6 hours at £22·60 per hr.
= 5 × 6 × 22·60 = £678·00 ✓
Time and a half pay:
2 days of 6 hours at
£22·60 × 1·5 per hr.
= 2 × 6 × 22·60 × 1·5
= £406·80 ✓
Gross Pay
= £678·00 + £406·80
= £1084·80 ✓

3 marks

Weekdays
- There are 5 weekdays and she works 6 hours on each of these days. This gives the factor 5 × 6
- Weekdays are paid at £22·60 per hour (basic rate)

Weekends
- There are 2 days at the weekend and she still works 6 hours on each day. This gives factor 2 × 6
- 'Time and a half' means she is paid her basic pay plus half again. So the rate is 1·5 × basic rate

Gross pay
- 'Gross pay' is the grand total of her pay before any deductions are made.

Q 4. *(a)*

	1	3	6	10
February	(F,1)	(F,3)	(F,6)	(F,10)
April	(A,1)	(A,3)	(A,6)	(A,10)
July	(J,1)	(J,3)	(J,6)	(J,10)
November	(N,1)	(N,3)	(N,6)	(N,10) ✓

1 mark

Table
- It is very important that no mistakes are made when completing the table. You will lose the mark if there is one mistake!
- Checking:
  1. Scan along each row, left to right, checking for the 1, 3, 6, 10 pattern.
  2. Scan down each column, top to bottom, checking for the F, A, J, N pattern

Q 4. *(b)*
Probability $= \frac{2}{16} = \frac{1}{8}$ ✓

1 mark

Probability
- Probability of an event $= \frac{\text{Number of outcomes that make the event happen (favourable outcomes)}}{\text{Total number of outcomes}}$
- In this case there are two 'favourable outcomes: (N, 1) and (N, 3) and there are 16 possible outcomes in total – namely all the entries in the table. So you have 2 out of 16 'favourable outcomes'
- The mark will be awarded for $\frac{2}{16}$, cancelling down to $\frac{1}{8}$ is not essential to gain the mark

See Int 2 Notes, Section 9, p. 35–36

Q 5.

Yes it is possible. Here is a suitable route:

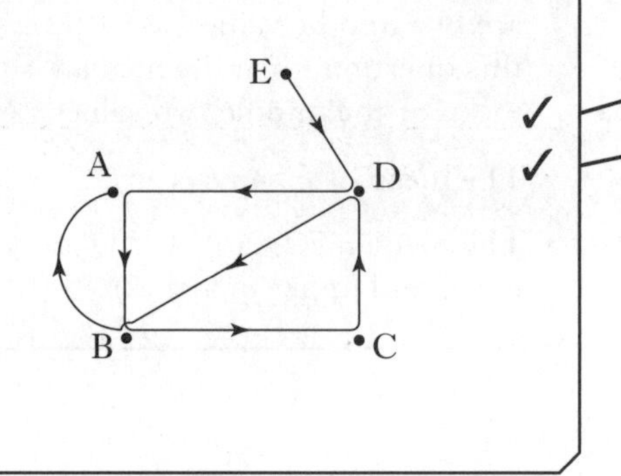

2 marks

Correct Route

- The route could also be described like this:

$E \rightarrow D \rightarrow A \rightarrow B \rightarrow C \rightarrow D \rightarrow B \rightarrow A$

- There are many other possible routes.
- Such a route is possible because there are exactly two nodes that are ODD namely E and A. E has 1 edge attached and A has 3 edges attached. Your route will start or finish at A or E. All other node are EVEN. You cannot start or finish at an EVEN node.

Q 6.

Draw CE. This radius meets chord AB at G.

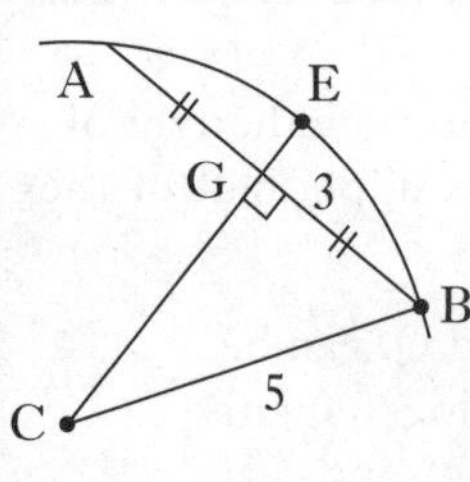

CE bisects AB so AG = GB = 3cm

CE is perpendicular to chord AB so angle CGB = 90° ✓

Use Pythagoras' Theorem in triangle CGB

$CG^2 = BC^2 - BG^2$
$= 5^2 - 3^2 = 25 - 9 = 16$

so $CG = \sqrt{16} = 4$ cm ✓

CE = 5 cm since CE is a radius so GE = CE − CG = 5 − 4 = 1cm

$\Rightarrow x = 1$ ✓

3 marks

Right-angled triangle

- There are several facts that you need to gather together to solve this problem:

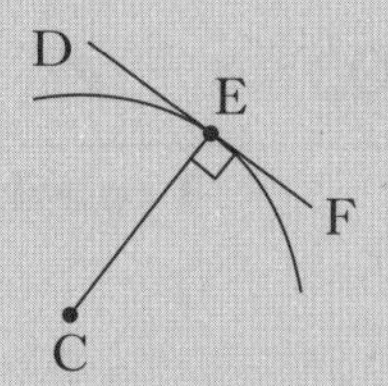

Fact 1 A tangent (DF) is perpendicular to the radius (CE) to the point of contact (E)

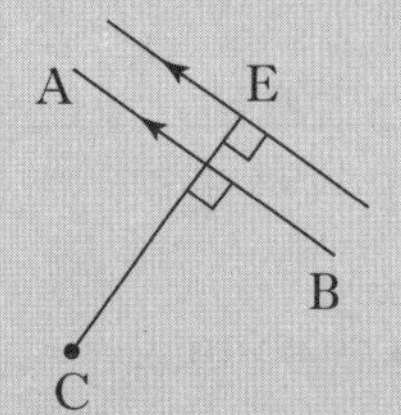

Fact 2 Any line parallel to DF will also be perpendicular to the radius CE

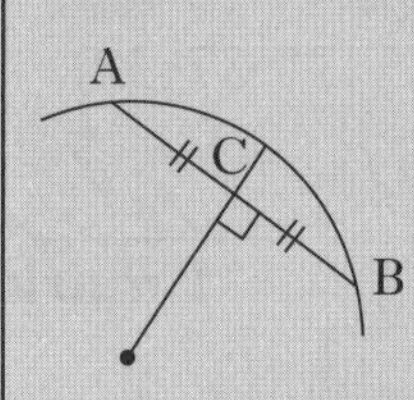

Fact 3 Any chord perpendicular to a radius is bisected by that radius. In this case G is the midpoint of chord AB.

All of these facts establish that triangle BCG is a right-angled triangle with GB = 3 cm and BC = 5 cm

Pythagoras' Theorem

- Correct use of Pythagoras' Theorem will gain you this strategy mark
- You should recognise this as a '3, 4, 5 triangle' in which case CG = 4 cm can just be written down: you will still gain this mark.

Calculation

- You have to recognise that GE is the required distance x. Notice that the radius CE is CG + GE = 4 + x so $5 = 4 + x$.

See Int 2 Notes, Section 5, p. 20

Q 7. *(a) (i)*
There are 18 measurements.
Divide them, in increasing order, into two groups of 9:
(5...............15)(18...............38)
↑
Median = $\frac{15+18}{2} = \frac{33}{2}$
So Median (Q_2) = 16·5 ✓

1 mark

Median
- To find the median the measurements are arranged in increasing order. The middle value is the median. If there are two middle values, as is the case in this question, then the median is the mean of the middle two values
- The mean of a and b is $\frac{a+b}{2}$
- The median is Q_2 (Q_1 and Q_3 are the lower and upper quartiles)

Q 7. *(a) (ii)*
Divide the lower 9 measurements, in increasing order, into two groups:
(5 6 7 7) 8 (9 12 12 15)
↑
The lower quartile (Q_1) = 8 ✓

1 mark

Lower Quartile
- Another way of thinking about the lower quartile is that it is the median of the lower half of the measurements.
- If you read the stem-and-leaf diagram from left to right and top row down to bottom row then you are reading the measurements in increasing order

Q 7. *(a) (iii)*
Divide the upper 9 measurements, in increasing order, into two groups:
(18 19 19 19) 23 (25 34 37 38)
↑
The upper quartile (Q_3) = 23 ✓

1 mark

Upper Quartile
- The upper quartile can be thought of as the median of the upper half of the measurements
- Although Q_1, Q_2 and Q_3 can be identified by counting along the numbers on the stem-and-leaf diagram it is safer to write out the measurements (9 in this case) so that no mistakes are made in counting along the measurements.

See Int 2 Notes, Section 8, p. 31

Q 7. *(a) (iv)*
Semi-interquartile range
$= \frac{1}{2}(Q_3 - Q_1)$
$= \frac{1}{2}(23 - 8) = \frac{1}{2} \times 15 = 7{\cdot}5$ ✓

1 mark

Semi-interquartile Range
- The formula $\frac{1}{2}(Q_3 - Q_1)$ is not given to you in the exam. You should memorise it!
- 'Semi' means 'half' and 'interquartile' means 'between the quartiles'. So the name 'semi-interquartile' helps you to remember the formula:

$\frac{1}{2}(Q_3 - Q_1)$
↗ ↖
'Semi' 'between the quartiles'.

See Int 2 Notes, Section 8, p. 32

Q 7. *(b)*
The times for the 2nd hive are less varied in their distribution about the median time. (semi-interquartile range is 3·5) than for the 1st hive (semi-interquartile range is 7·5 which is greater than 3·5) ✓

1 mark

Statement
- A larger semi-interquartile range means the measurements are more spread out and a smaller semi-interquartile range means the measurements are less spread out.
- It is important that you are able to communicate that you know that hive 1 times are more spread out because the semi-interquartile range, 7·5, is greater than the semi-interquartile range, 3·5, for hive 2. This comparison is crucial for gaining this mark.

See Int 2 Notes, Section 8, p. 33

Q 8. *(a)*
$a = 2$, $s = 3{\cdot}8$, $h = 7$
$A = 5as + 5sh$
$= 5 \times 2 \times 3{\cdot}8 + 5 \times 3{\cdot}8 \times 7$ ✓
$= 10 \times 3{\cdot}8 + 19 \times 7$
$= 38 + 133 = 171$ ✓

2 marks

Substitution
- The 1st mark is for the correct substitution i.e. correctly replacing each of the letters with their values
- Remember to double-check the values you have used. Sometimes values can be copied wrongly. It is best to copy them at the top of your working as has been done in these solutions.

Calculation
- With no calculator in Paper 1 you are looking for more efficient calculation methods eg you choose $5 \times 2 = 10$, multiplying by 10 is easier.

Q 8. *(b)*
$A = 270$, $a = 3$, $h = 6$
$A = 5as + 5sh$
$\Rightarrow 270 = 5 \times 3 \times s + 5 \times s \times 6$ ✓
$\Rightarrow 270 = 15s + 30s$ ✓
$\Rightarrow 270 = 45s \Rightarrow s = \frac{270}{45} = 6$ ✓

3 marks

Substitution
- The 1st mark is for correct substitution

Start Solving
- You recognise an equation and so attempt to simplify the right hand side

Solution
- Divide both sides of the equation by 45
- Notice $\frac{270}{45} = \frac{9 \times 30}{9 \times 5} = \frac{9 \times 5 \times 6}{9 \times 5}$

the 9 and 5 factors can be cancelled.

Q 9. *(a)*

Mark	Frequency	Comulative Frequency
1 – 10	7	7
11 – 20	12	19
21 – 30	19	38
31 – 40	24	62
41 – 50	20	82
51 – 60	9	91
61 – 70	7	98
71 – 80	2	100 ✓

1 mark

Table
- This mark is for completing the last column of the table correctly
- You are adding each new frequency number as you go:

$19 = 7 + 12$, $38 = 19 + 19$, $62 = 38 + 24$ etc

- The total of the frequency column should equal the last entry (100) in the cumulative frequency column.

Q 9. *(b)*

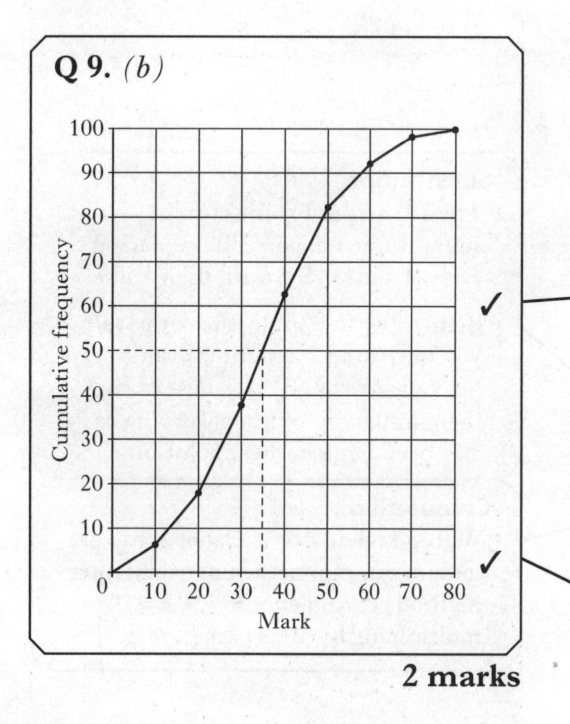

2 marks

Scales/Labels
- Cumulative frequency is always on the vertical axis, in this case 0 to 100 choose a scale that gives a reasonable size for your diagram. Don't make your diagram too small
- Marks range from 1 to 80 but always start your scale from the number that would give the upper bound of the grouping previous to the first grouping — in this case it is 0.
- Remember to label your two scales in this case 'Cumulative frequency' and 'Mark'

Points
- Use the upper bounds of the intervals — in this case 10, 20, 30 etc. So the first point is (10, 7) then (20, 12) etc.

Q 9. *(c)*

Median mark ≑ 35 ✓

1 mark

Median
- Half the last comulative frequency total is 50. Draw a horizontal line to the curve then drop a vertical to the 'x-axis and read off the median (35)

WORKED ANSWERS: EXAM A — PAPER 2

Q 1. *(a)*

$x + y = 22$ ✓

1 mark

1st equation
- There is information in the question that is not needed for this equation: £80 and £120.
- The total number of rooms is 22 made up of x singles and y doubles

Q 1. *(b)*

$80x + 120y = 2320$ ✓

1 mark

2nd equation
- This is the 'cost equation': x rooms at £80 and y rooms at £120 giving a total income of £2320

See Int 2 Notes, Section 7, p. 25

Q 1. *(c)*

$80x + 120y = 2320 \; (\div 40)$

$\Rightarrow 2x + 3y = 58$

Solve:

$$\left.\begin{aligned} x+y&=22 \\ 2x+3y&=58 \end{aligned}\right\} \begin{aligned} &\times 2 \rightarrow 2x+2y=44 \\ &\rightarrow \underline{2x+3y=58} \end{aligned} \quad ✓$$

Subtract: $-y = -14$

$\Rightarrow y = 14$ ✓

Now substitute $y = 14$ in $x + y = 22 \Rightarrow x + 14 = 22 \Rightarrow x = 8$ ✓

So there are 8 single rooms and 14 double rooms ✓

4 marks

Strategy
- An attempt to use simultaneous equations will gain you this strategy mark

Values
- If you follow the correct method in your working, even making calculation errors, you can still gain this mark
- Notice that dividing by 40 greatly simplifies the 2nd equation making all the subsequent working much easier

Correct values
- Producing the correct values of x and y with the relevant working will gain you this mark.

Statement
- You have to go beyond the values of x and y by making a statement concerning the number of each type of room that the hotel has.

See Int 2 Notes, Section 7, p. 26–27

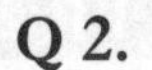

Q 2.

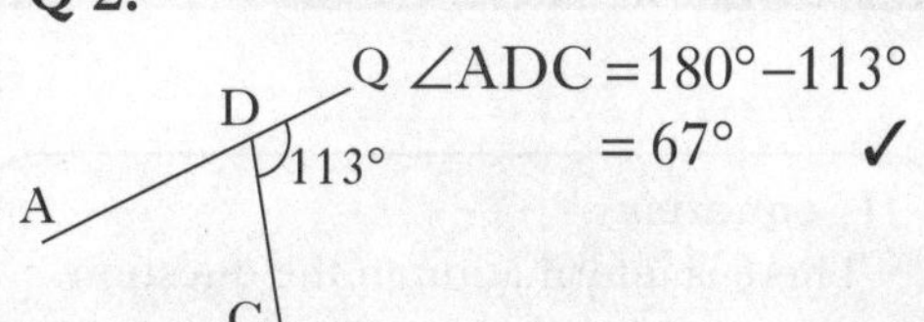

$\angle ADC = 180° - 113°$
$= 67°$ ✓

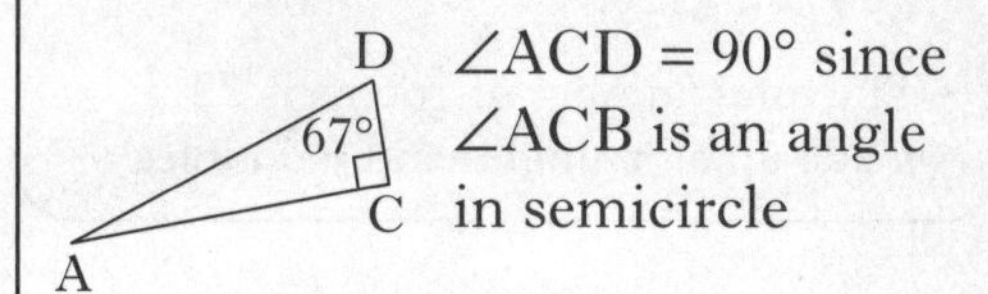

$\angle ACD = 90°$ since $\angle ACB$ is an angle in semicircle

so $\angle DAC = 90° - 67° = 23°$ ✓

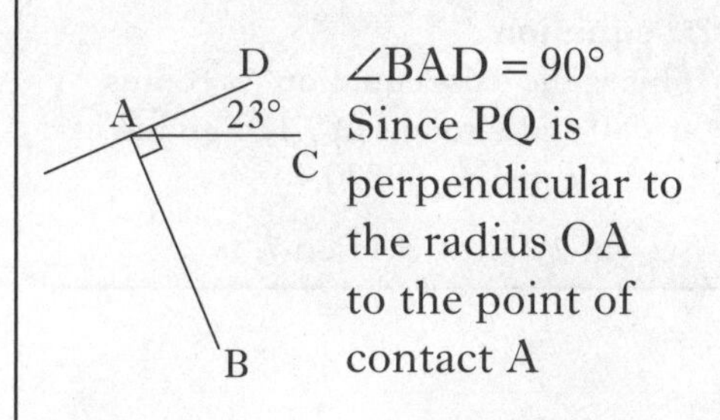

$\angle BAD = 90°$ Since PQ is perpendicular to the radius OA to the point of contact A

so $\angle BAC = 90° - 23° = 67°$ ✓

3 marks

Angle ADC

- 'straight angles' add to 180°. In this case ADQ is a 'straight angle' so $\angle ADC + \angle QDC = 180°$

Angle DAC

- AB is a diameter with C on the circumference so $\angle ACB$ is an 'angle in a semicircle'. All such angles are 90°
- DCB is a straight line so $\angle DCA + \angle ACB = 180°$

Angle BAC

- The tangent (PQ) is perpendicular to a radius (OA) to the point of contact (A) so $\angle DAB = 90°$ giving $\angle DAC + \angle CAB = 90°$

See Int 2 Notes, Section 5, p. 19–20

Q3

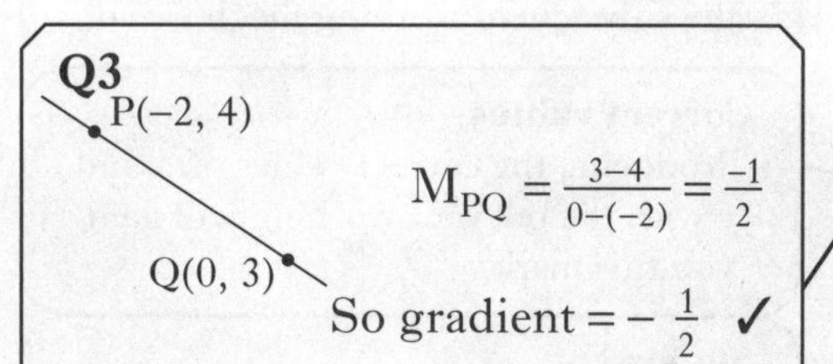

$$M_{PQ} = \frac{3-4}{0-(-2)} = \frac{-1}{2}$$

So gradient $= -\frac{1}{2}$ ✓

The y-intercept is (0, 3)

So the equation of the line is:

$y = -\frac{1}{2}x + 3$ (×2) ✓

$\Rightarrow 2y = -x + 6$ ✓

$\Rightarrow 2y + x = 6$

3 marks

Gradient

- The formula used here is:

$$\text{gradient} = \frac{\text{distance up or down } (y\text{-difference})}{\text{distance along } (x\text{-difference})}$$

- For the points $A(x_1, y_1)$ and $B(x_2, y_2)$

gradient of AB $(m_{AB}) = \dfrac{y_2 - y_1}{x_2 - x_1}$ ← y-difference ← x-difference

***y*-intercept**

- The y-intercept (0,3) gives $c = 3$ in the formula: $y = mx + c$

Equation

- Equation of a straight line:

$y = mx + c$

↗ gradient ↑ y-intercept (0,c)

- In this case $m = \frac{1}{2}$ and $c = 3$
- $y = -\frac{1}{2}x + 3$ will gain you this 3rd mark.

Removing the fraction and rearranging to $2y + x = 6$ is not essential, but this is a good form for questions where simultaneous equations are then required.

See Int 2 Notes, Section 3, p. 9–11

Q 4.

	Frequency	Fraction	Angle	
A:	27	$\frac{27}{81}=\frac{3}{9}=\frac{1}{3}$	$\frac{1}{3}$ of 360° = 120°	✓
B:	9	$\frac{9}{81}=\frac{1}{9}$	$\frac{1}{9}$ of 360° = 40°	
C:	27	$\frac{27}{81}=\frac{3}{9}=\frac{1}{3}$	$\frac{1}{3}$ of 360° = 120°	
D:	18	$\frac{18}{81}=\frac{2}{9}$	$\frac{2}{9}$ of 360° = 80°	✓
	Total = 81		Total = 360°	

Pie chart showing 'cleaner preference'

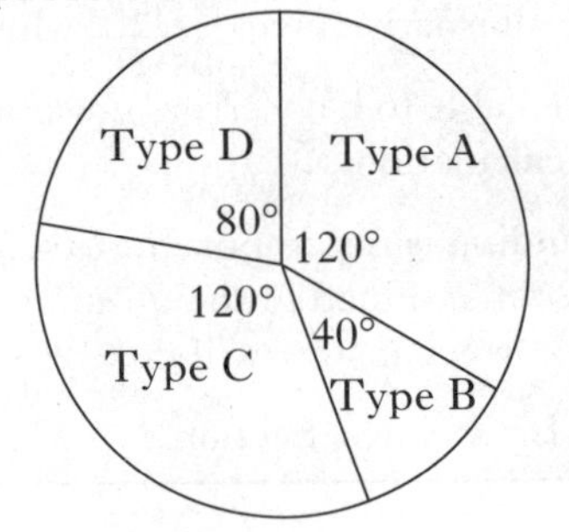

✓

3 marks

Fractions of 360°

- The fraction $\frac{\text{frequency}}{\text{total}}$ gives the size of each sector of the pie chart. You find this fraction of 360° to give you the sector angle
- This mark is for knowing the method for calculation of the sector angles

Angles

- Calculation of the correct sector angles will gain you this mark
- Always check that your sector angles (in this case 120°, 40°, 120°, 80°) add up to 360°!

Pie chart

- All sectors should be labelled eg 'Type A' etc
- You should be very accurate drawing your Pie Chart as you are only allowed up to a 2° error in each angle

See Int 2 Notes, Section 8, p. 29

Q5. *(a)*

Semicircle

Rectangle

radius of semicircle
$= \frac{1}{2} \times 40 = 20$ cm

total area
= area of rectangle + area of semicircle
= length × breadth $+ \frac{1}{2}\pi r^2$ ✓
$= 70 \times 40 + \frac{1}{2} \times \pi \times 20 \times 20$
= 2800 + 628·31...
= 3428·31... cm²

Volume = Area of end × depth ✓
= 3428·31 × 35
= 119991·14... ✓
≑ 120000 cm³ ✓
(to 2 significant figures)

4 marks

Substitution

- Correct substitution of length = 70, breadth = 40 and radius = 20 into the formulae will gain you this mark
- A common error is using the diameter value (40 cm) when you should be using the radius value (20 cm). The area of a circle formula $A = \pi r^2$ uses the radius value.
- Radius = $\frac{1}{2}$ × Diameter

Strategy

- Knowing to find the area of the end of this prism and multiply by the depth will gain you the strategy mark

Calculation

- Never round answers until the end of a long calculation like this

Rounding

- A mark is allocated for correctly rounding your answer to 2 significant figures.
- This mark can be gained even if your answer is wrong before the rounding. The mark is for knowing how to round to 2 sig figs.

Q 5. *(b)*

Volume of drawer = Area of quarter circle × length ✓

$\Rightarrow 10000 = \frac{1}{4}\pi r^2 \times \text{length}$ ✓

$\Rightarrow 10000 = \frac{1}{4}\pi \times 20^2 \times \text{length}$

$\Rightarrow 10000 = 314{\cdot}15\ldots \times \text{length}$

$\Rightarrow \frac{10000}{314{\cdot}15\ldots} = \text{length}$

so length = 31·83...

≑ 31·8 cm ✓

(to 1 decimal place)

3 marks

Strategy
- Knowing how to find the volume of the drawer will gain you this mark

Equation
- This time you know the volume so there is an equation to solve. You will gain this mark for correctly setting up the equation

Solving
- It is easier to calculate $\frac{1}{4}\pi \times 20^2$ first and then dividing through by the answer 314·15...

  The alternative gives $\frac{4 \times 10000}{\pi \times 20^2}$ which is more liable to have errors creep into the calculation
- Rounding is not important here as it is not mentioned in the question - so any correct answer will do.

See Int 2 Notes, Section 2, p. 8

Q 6.

Circumference of Inside Circle = π × D ✓

where D = 2 × 8 = 16 cm

So Circumference = π × 16 = 16π cm

Fraction of Circumference = $\frac{35}{60} = \frac{7}{12}$ ✓

Distance travelled = $\frac{7}{12} \times 16\pi$ ✓

= 29·32... ≑ 29·3 cm

(to 3 sig figs) ✓

4 marks

Strategy
- Knowing how to find the circumference will gain you this mark
- Either C = πD or C = 2π*r* (*r* = 8)

Fraction
- A complete turn i.e whole circumference is travelled in 60 minutes. Tip has travelled for 35 minutes out of 60 minutes i.e. $^{35}/_{60}$

Strategy
- To gain this mark requires evidence that you knew to find $\frac{7}{12}$ of the circumference

Calculation
- On calculator:

  7 ÷ 1 2 × 1 6 × π EXE
- Any correct rounding is acceptable here

See Int 2 Notes, Section 5, p. 17–18

Q 7. *(a) (i)*

Mean = $\frac{54+45+51+50+48+53+49}{7}$

= $\frac{350}{7}$ = 50 ✓

1 mark

Mean
- Mean = $\frac{\text{sum of the numbers}}{\text{number of numbers}} = \frac{\Sigma x}{n} = \overline{x}$

Q 7. *(a) (ii)*

x	$x-\bar{x}$	$(x-\bar{x})^2$
54	4	16
45	−5	25
51	1	1
50	0	0
48	−2	4
53	3	9
49	−1	1

$\Sigma(x-\bar{x})^2 = 56$ ✓

$s = \sqrt{\frac{\Sigma(x-\bar{x})^2}{n-1}} = \sqrt{\frac{56}{6}} \doteq 3{\cdot}1$ (to 1 dec. pl.) ✓ ✓

3 marks

Squared deviations

- This mark is gained for the correct last column in the table: the squared values of the deviations from the mean

Substitution

- The formula $s = \sqrt{\frac{\Sigma(x-\bar{x})^2}{n-1}}$ is given to you in the exam on your formulae page
- This mark is for correctly substituting $\Sigma(x-\bar{x})^2 = 56$ and $n-1=6$ into the formula

Calculation

- If you know how to use the STAT mode on your calculator to check $s = 3{\cdot}1$ then you should do this. Remember just writing the answer down from the calculator with no working will not gain you the marks

Q 7. *(b)*

On average there are more matches in Machine B's boxes (mean = 52 is greater than mean = 50 for machine A) ✓

There is much more variation in the number of matches in Machine A's boxes than those from Machine B (s = 3·1 for Machine A is greater than s = 1·6 for Machine B) ✓

2 marks

1st statement

- This concerns the 'average' contents. Back up your statement with the statistics (52 and 50)

2nd statement

- This concerns the distribution about the mean. A larger standard deviation means more variation about the mean.

See Int 2 Notes, Section 9, p. 33–34

Q 8. *(a)*

Earnings

= £500 + 2·5% of £2400 ✓

= £500 + £60

= £560 ✓

2 marks

Strategy

- Knowing to find 2·5% and then add the result to £500 will gain you this 1st mark.
- The basic wage has nothing to do with the commission — no sales: she will still earn her basic wage of £500.

Calculation

- The calculator calculation is:

500 + 2·5 ÷ 100 × 2400

Q 8. *(b)*

Commission

= £710 – £500 = £210 ✓

Test:

2·5% of £2500 = £62·50

2·75% of £5000 = £137·50 ✓

since £210 is greater than £137·50

her commission rate is 3% ✓

3% of sales = £210

so 1% of sales = £70

so 100% of sales = £70 × 100

= £7000 ✓

4 marks

Commission

- The extra over and above £500 is commission

Strategy

- To find the rate that applies you need to experiment to find the greatest for each band: £62·50 is not enough, £137·50 is not enough so only the 3% band will give £210.

Statement

- You must declare clearly that you have found that 3% is the rate that applies.

Calculation

- This is a 'proportion' problem. If you know 3% then divide by 3 to find 1%. The whole amount is 100% so now multiply by 100.

Q 9. *(a)*

Use the Cosine Rule in triangle ABC ✓

$$\text{Cos C} = \frac{a^2 + b^2 - c^2}{2ab}$$

$$\Rightarrow \text{Cos C} = \frac{40^2 + 60^2 - 35^2}{2 \times 40 \times 60} ✓$$

= 0·828...

So angle C = $\cos^{-1}$ (0·828...)

= 34·09...

≑ 34·1° ✓

3 marks

Strategy

- Knowing to use the Cosine Rule will gain you this mark

Substitution

- The formula given to you in the exam on the formulae page is: $\cos A = \frac{b^2 + c^2 - a^2}{2bc}$.

  You should practise changing this formula to the forms: $\cos B = \frac{a^2 + c^2 - b^2}{2ac}$ and $\cos C = \frac{a^2 + b^2 - c^2}{2ac}$

- When you know the 3 sides of a triangle and are asked to find an angle then you should know to use the Cosine Rule.

Calculation

- Your calculator must be set in DEG mode. There should be a 'D' or 'DEG' on your display otherwise using $\boxed{\cos^{-1}}$ will give you the wrong answer.

See Int 2 Notes, Section 6, p. 24

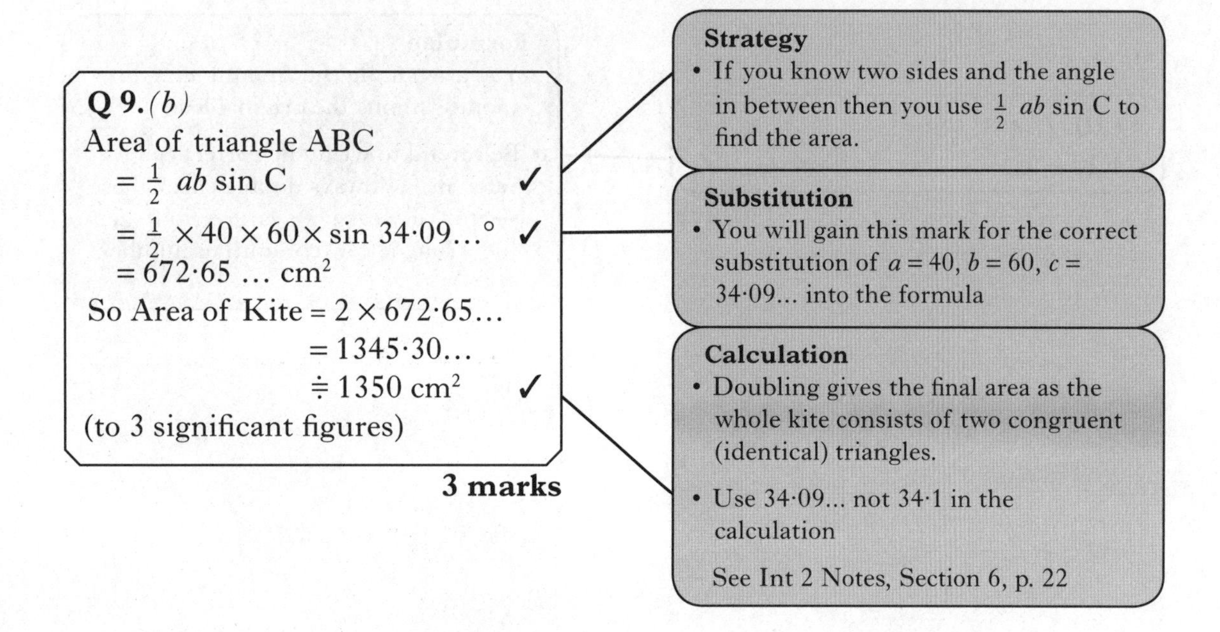

Q 9. *(b)*

Area of triangle ABC

$= \frac{1}{2}\ ab \sin C$ ✓

$= \frac{1}{2} \times 40 \times 60 \times \sin 34{\cdot}09\ldots^\circ$ ✓

$= 672{\cdot}65\ \ldots\ \text{cm}^2$

So Area of Kite $= 2 \times 672{\cdot}65\ldots$

$= 1345{\cdot}30\ldots$

$\doteqdot 1350\ \text{cm}^2$ ✓

(to 3 significant figures)

3 marks

Strategy
- If you know two sides and the angle in between then you use $\frac{1}{2}\ ab \sin C$ to find the area.

Substitution
- You will gain this mark for the correct substitution of $a = 40$, $b = 60$, $c = 34{\cdot}09\ldots$ into the formula

Calculation
- Doubling gives the final area as the whole kite consists of two congruent (identical) triangles.
- Use 34·09... not 34·1 in the calculation

See Int 2 Notes, Section 6, p. 22

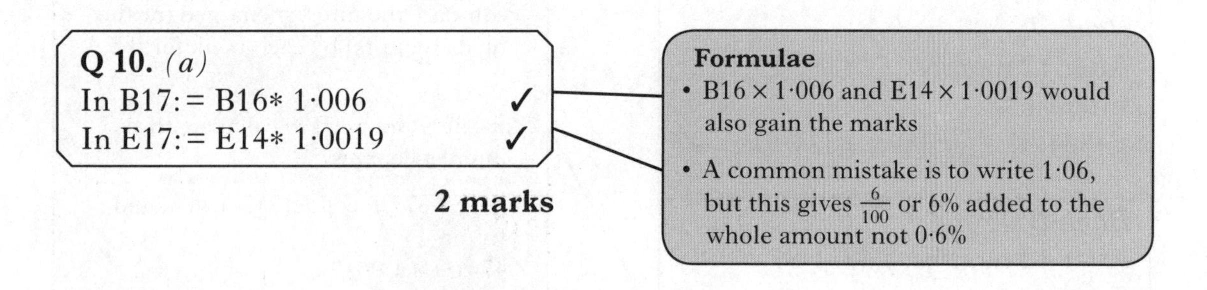

Q 10. *(a)*

In B17: = B16* 1·006 ✓

In E17: = E14* 1·0019 ✓

2 marks

Formulae
- B16 × 1·006 and E14 × 1·0019 would also gain the marks
- A common mistake is to write 1·06, but this gives $\frac{6}{100}$ or 6% added to the whole amount not 0·6%

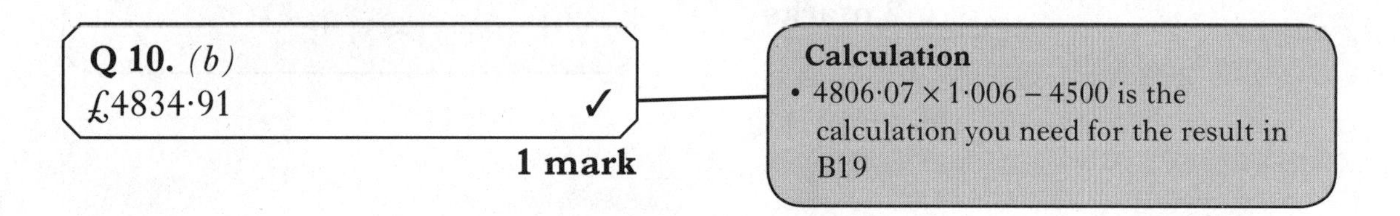

Q 10. *(b)*

£4834·91 ✓

1 mark

Calculation
- 4806·07 × 1·006 – 4500 is the calculation you need for the result in B19

Q 10. *(c)*
=E17 – E5 ✓

1 mark

Formulae
- You are finding the amount after 12 months minus the original loan
- Be careful to write the correct cell references. Always double check these — it is very easy in a large grid to get the wrong letter (column) or number (row)

Q 10. *(d)*
Bank A
After 12 months: £3834·91
Interest =
£4834·91 –£4500 = £334·91 ✓

Bank B
After 12 months: £4851·87
Interest =
£4851·87 – £4500 = £351·87 ✓
The loan from Bank B is more expensive by:
£351·87 – £334·91 = £16·96 ✓

2 marks

Strategy
- To answer this question you have to compare the interest charged (or the final amounts) by each bank for the loan.

You therefore need to calculate these interest charges

- £4851·87 (in cell E17) is calculated from:
  4761·4 × 1·0019

This is the formula you gave in part (a)

Statement
- A clear statement is required.
- Alternatively the amounts could be subtracted:

  £4851·87 – £4834·91

WORKED ANSWERS: EXAM B — PAPER 1

Q 1.

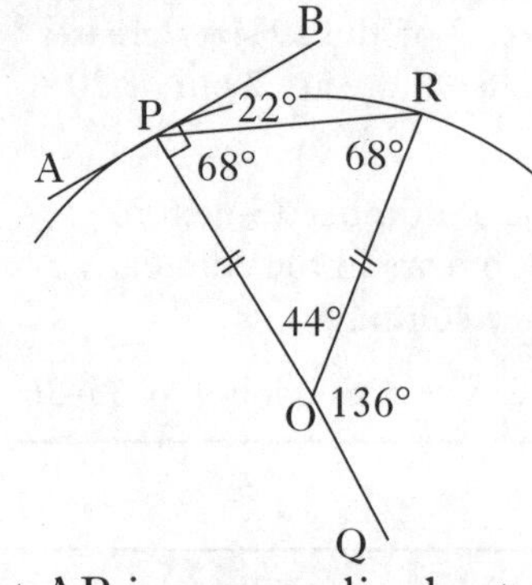

Tangent AB is perpendicular to the radius OP

So ∠OPR = 90°–22° = 68° ✓

Triangle OPR is isosceles (OP and OR are equal radii)

So ∠PRO = ∠OPR = 68°

In triangle OPR:

∠POR = 180°– (68° + 68°) = 44° ✓

⇒ ∠QOR = 180° – 44° = 136° ✓

3 marks

Tangent/Radius

- A tangent to a circle is perpendicular to the radius to the point of contact. In this case ∠OPB = 90°

Isosceles Triangle

- Any two radii in a circle are equal in length
- The angles opposite the equal sides in an isosceles triangle are themselves equal. In this case ∠OPR and ∠ORP
- The three angles in triangle OPR add up to 180°

So ∠OPR + ∠ORP + ∠POR = 180°

giving 68° + 68° + ∠POR = 180°

So ∠POR = 180° – 68° – 68° = 44°

Straight Angle

- POQ is a straight line (a diameter) so ∠POR + ∠ROQ = 180°
  ⇒ 44° + ∠ROQ = 180°
  giving ∠ROQ = 180° – 44° = 136°
- You do not need to give all the reasons in your solution to gain the marks.

See Int 2 Notes, Section 5, p. 19

Q 2. *(a)*

Size	tally	frequency	cumulative frequency ✓
6	|	1	1 ✓
6½	||	2	3
7	|||	3	6
7½	~~||||~~ ||	7	13
8	||||	4	17
8½	|	1	18
9	||	2	20
		Total = 20	

2 marks

frequency

- The 1st mark is for a correct frequency column
- It is useful to check your frequency column with a total. This should equal the number of numbers.

Cumulative frequency

- The 2nd mark is for the correct cumulative frequency column
- Add the next frequency to the previous cumulative frequency to complete this column
- Your final value (20) should equal the frequency total, if not, you have made a calculation error.

See Int 2 Notes, Section 8, p. 30

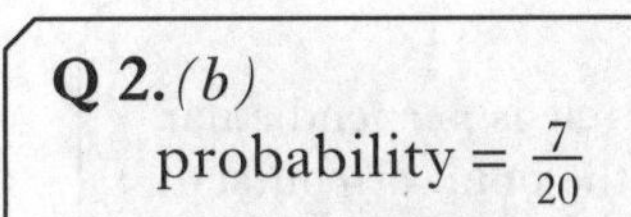

Q 2. *(b)*

probability = $\frac{7}{20}$ ✓

1 mark

Probability

- Probability = $\dfrac{\text{Number of favourable outcomes}}{\text{Total number of outcomes}}$

  In this case since there are 13 sizes that are equal to or less than 7½ then this leaves 7 of the 20 sizes greater than 7½. So there are 7 out of 20 giving $^{7}/_{20}$

- The crucial number 13 next to size 7½ appears in the cumulative frequency column.

See Int 2 Notes, Section 9, p. 35–36

Q 3. *(a)*

Dotplot of Ages for oil rig Alpha

age (years)	19	20	21	22	23	24	25	26	27	28	29
dots	•		••••	••	•••••••	••			•		•

✓ ✓

2 marks

Dotplot

- You should first determine the least (19) and greatest (29) of the ages as this will allow you to plan the scale.
- You have access to squared paper during your exam and you should use it for a question of this type
- Work systematically through the data placing a new dot on your dotplot for each new piece of data.

Completion

- Count the number of dots when you have completed your dotplot. In this case there should be 21, the same number as the pieces of data.
- Remember to give your dotplot a title and also to label the axis to give a meaning to the numbers

See Int 2 Notes, Section 8, p. 30

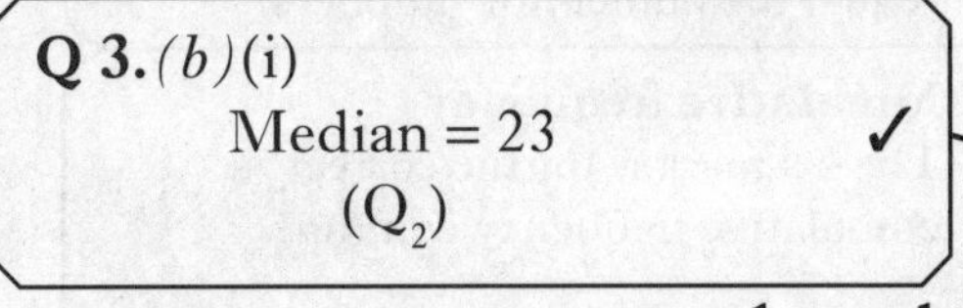

Q 3. *(b)*(i)

Median = 23 ✓
(Q_2)

1 mark

Median

- The symbol Q_2 is used for the median
- For 21 ages: (1st… 10th) (11th) (12th… 21st). So when placed in increasing order the 11th age gives the median age for this data i.e 23

Q 3. *(b)*(ii)

Lower Quartile = $\frac{21+22}{2} = 21{\cdot}5$ ✓
(Q_1)

1 mark

Lower Quartile

- The symbol Q_1 is used for the lower quartile
- This is the median of the lower half of the ages i.e. the median of the 1st ten of the ages in increasing order: (1st… 5th) (6th… 10th). It is the mean of the 5th and 6th ages i.e. the mean of 21 and 22

Q 3. *(b)*(iii)

Upper Quartile $= \frac{23+24}{2} = 23{\cdot}5$ ✓
(Q_3)

1 mark

Upper Quartile

- The Symbol Q_3 is used for the upper quartile
- This is the median of the upper half of the ages i.e. the median of the 2nd group of 10 ages in increasing order: (12th… 16th) (17th… 21st). It is the mean of the 16th and 17th ages i.e. the mean of 23 and 24

See Int 2 Notes, Section 8, p. 31

Q 3. *(c)*

For oil rig Alpha:
semi-interquartile range
$= \frac{1}{2}(23{\cdot}5 - 21{\cdot}5) = 1$ ✓
The age distribution is more spread out (about the mean age) on oil rig Beta compared to that on oil rig Alpha since the semi-interquartile range of 3·5 for Beta is greater than the 1 for Alpha. ✓

2 marks

Semi-interquartile range

- You need to calculate the statistic: Semi-interquartile range $= \frac{1}{2}(Q_3 - Q_1)$ for oil rig Alpha and then compare it to the same statistic for oil rig Beta

Comparison

- It is important you back any comparison statement with the relevant statistics. In this case the semi-interquartile range: the greater the range the more spread out the data is around the mean

See Int 2 Notes, Section 9, p. 32–33

Q 4

gradient $= \frac{3}{2}$ ✓
y-intercept is $(0, -3)$ ✓
equation is $y = \frac{3}{2}x - 3$ ✓
$(\Rightarrow 2y = 3x - 6 \Rightarrow 2y - 3x = -6)$

3 marks

gradient

- You must take great care over the scales on each axis. On the x-axis 1 square = 1 unit but on the y axis 1 square = 3 units. It appears the gradient should be $\frac{1}{2}$ using the definition:

$$\text{gradient} = \frac{\text{distance up (or down)}}{\text{distance along}}$$

however in this case the distance up is 3 units not 1 unit. The gradient is $\frac{3}{2}$

***y*-intercept**

- The y-intercept is where the line crosses the y-axis and this determines the value of c in "$y = mx + c$"
- Careful with the y-axis scale. The y-intercept is 1 square down which is 3 units down so $c = -3$

Equation

- You use the equation:

$$y = mx + c$$

gradient ↗ (pointing to m)   ↖ $(0, c)$ is the y-interept

with $m = \frac{3}{2}$ and $c = -3$

- "$y = \frac{3}{2}x - 3$" will gain you this 3rd mark. The further rearrangement given in the solution is not essential.

See Int 2 Notes, Section 3, p. 9–11

Q 5. *(a)*

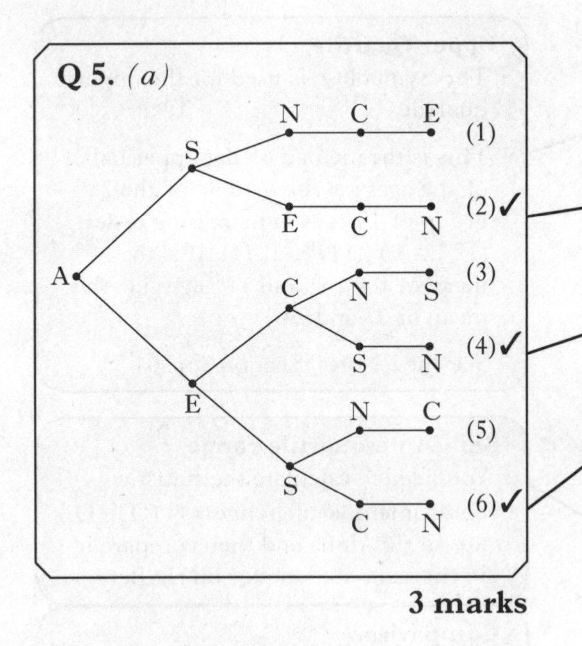

3 marks

Diagram
- The 1st mark is for 2 correct routes. The 2nd mark is for a further 2 correct routes
- The letters are not confusing since all 5 town names begin with different letters
- The routes are numbered (1) to (6) to help explain part (6)

Q 5. *(b)*

The totals for the 6 routes are: ✓
(1) 15 + 17 + 18 + 24 = 74
(2) 15 + 19 + 24 + 18 = 76
(3) 9 + 24 + 18 + 17 = 68
(4) 9 + 24 + 16 + 17 = 66 ✓
(5) 9 + 19 + 17 + 18 = 63 ✓
(6) 9 + 19 + 16 + 18 = 62
The shortest of these is (6) at 62 km:
$A \rightarrow E \rightarrow S \rightarrow C \rightarrow N$

3 marks

Strategy
- You should know to calculate totals for each route

Shortest Route
- All your working is essential
- You might find it easier to put each route length into the tree diagram
- A clear statement about the shortest route is necessary

Q 6. *(a)*
Long vertical rods $= (x + 4)$ cm (×2)
Short vertical rods $= x$ cm (×4)
Horizontal rods $= (x - 2)$ cm (×5)
Total length ✓
$= 2(x + 4) + 4x + 5(x - 2)$
$= 2x + 8 + 4x + 5x - 10$
$= 11x - 2$ cm ✓
so $L = 11x - 2$

2 marks

Strategy
- Identify the three different lengths of rod and the number of each type:

  2 rods of length $(x + 4)$ cm: $2(x + 4)$ cm

  4 rods of length x cm: $4x$ cm

  and 5 rods of length $(x - 2)$ cm: $5(x - 2)$ cm

Simplification
- Since $11x - 2$ is given in the question it is very important that you show all your working clearly.

Q 6. *(b)*
$L = 11x - 2 = 713$
$\Rightarrow 11x = 715$
$\Rightarrow x = \frac{715}{11} = 65$ ✓

1 mark

Value of x
- You should set up an equation based on the information you are given. The total length of rod is 713 cm but previously you have determined that an expression for this total is $11x - 2$. You now 'equate' these: $11x - 2 = 713$

Q 7. *(a)*
$n = 5$, $a = 10$, $S = 0{\cdot}5$

$$L = \frac{Sa\,(n-1)}{n} = \frac{0{\cdot}5 \times 10 \times (5-1)}{5}$$ ✓

$$= \frac{5 \times 4}{5} = 4$$ ✓

2 marks

Substitution
- The 1st mark is for replacing the letters by their values
- Remember that letters side by side means multiply eg Sa means $S \times a$

Calculation
- Calculations inside brackets are carried out first so $5 - 1 = 4$

Q 7. *(b)*
$L = 50$, $a = 3$, $S = 20$

$$L = \frac{Sa\,(n-1)}{n}$$

$$\Rightarrow 50 = \frac{20 \times 3 \times (n-1)}{n}$$ ✓

$\Rightarrow 50\,n = 60\,(n - 1)$
$\Rightarrow 50\,n = 60n - 60$ ✓
$\Rightarrow 60 = 60n - 50n$
$\Rightarrow 60 = 10n \Rightarrow n = 6$ ✓

3 marks

Substitution
- Correct substitution of the values into the formula will gain this 1st mark.

Start Solving
- Multiply both sides by n to get rid of the fraction
- Get rid of the brackets by multiplying out

Solution
- The variable (letter) n should be gathered onto one side of the equals sign.
- Add 60 to both sides, subtract $50n$ from both sides and then divide both sides by 10

WORKED ANSWERS: EXAM B **PAPER 2**

Q 1. *(a)*

$2x^2 + x - 28$ ✓

$= (2x - 7)(x + 4)$ ✓

2 marks

1st factor

- One correct factor will gain you this 1st mark

2nd factor

- when you have completed your factorisation you should check your answer by multiplying out using 'FOIL'
- Most mistakes are to do with the positive and negative signs. Check these carefully

See Int 2 Notes, Section 4, p. 16

Q 1. *(b)*

$(2a - 3b)(3a + 2b) + 2ab$ ✓

$= 6a^2 + 4ab - 9ab - 6b^2 + 2ab$ ✓

$= 6a^2 - 3ab - 6b^2$ ✓

3 marks

Starting removing the brackets

- 2 correct terms will gain you this mark.
- The pattern used for multiplying out the brackets is: Firsts, Outsides, Insides, Lasts:

F L I O

$(2a - 3b)(3a + 2b)$

Firsts	Outsides	Insides	Lasts
$2a \times 3a$	$2a \times 2b$	$-3b \times 3a$	$-3b \times 2b$
$6a^2$	$4ab$	$-9ab$	$-6b^2$

Completing removing the brackets

- There are 4 terms with the Outsides (4ab) and Insides (–9ab) being 'like terms'

Like terms

- In this case there are 3 'like terms' namely 4ab, –9ab and 2ab which combine to give –3ab

See Int 2 Notes, Section 4, p. 12–14

Q 2.

A rise of 28·5% means there will be 128·5% the next day

Multiplication factor = 1·285 ✓

Cases on Thursday

$= 6400 \times 1{\cdot}285^3$ ✓

$= 13579{\cdot}6\ldots$

$\doteqdot 13600$ ✓

(to 3 significant figures)

3 marks

Multiplication factor

- Starting with 6400 cases you consider this as 100% of the cases. Increasing by 28·5% will therefore give you 128·5% or $\frac{128{\cdot}5}{100} = 1{\cdot}285$.
- Each increase of 28·5% is given by a multiplication by 1·285. So three consecutive increases will be given by three such multiplications

Strategy

- $\times\, 1{\cdot}285 \times 1{\cdot}285 \times 1{\cdot}285$ is the same as $\times\, 1{\cdot}285^3$
- Here is a useful diagram:

Monday (6400) → Tuesday → Wednesday → Thursday

× 1·285 × 1·285 × 1·285

Calculation

- Rounding correctly is important as it is mentioned in the question, so follow the instructions!

See Int 2 Notes, Section 1, p. 5

Q 3.

Six sectors would fit to make a complete circle

20 m, Arc, 20 m

Arc length ✓

$= \frac{1}{6} \times \pi D$ ✓

$= \frac{1}{6} \times \pi \times 40$

$= 20{\cdot}94\ldots.$

Total length of fencing

$= 20 + 20 + 20{\cdot}94\ldots.$

$= 60{\cdot}94\ldots.$

$\doteqdot 60{\cdot}9$ m (to 1 dec pl.) ✓

3 marks

Circumference

- Your first step is to find the circumference of the whole circle using $C = \pi D$
- None of the circle formulae are given to you on your formulae page so you must learn them!

Arc length

- This mark is awarded for knowing how to find the arc length, it is $\frac{1}{6}$ of the circumference.

Perimeter

- The two radii are part of the fencing of Pen 1 and so must be added to the arc length
- Since accuracy is not mentioned in the question any correct rounding for the answer will do.

See Int 2 Notes, Section 4, p. 17–18

Q 4. *(a)*

Mean $= \frac{8+12+13+12+11+11+10}{7} = \frac{77}{7}$ ✓

$= 11°C$

x	$x-\overline{x}$	$(x-\overline{x})^2$
8	−3	9
12	1	1
13	2	4
12	1	1
11	0	0
11	0	0
10	−1	1

✓

$\Sigma(x-\overline{x})^2 = 16$

✓

$s = \sqrt{\frac{\Sigma(x-\overline{x})^2}{n-1}} = \sqrt{\frac{16}{6}} = \sqrt{2.66}$ ....

$= 1{\cdot}63.... \doteq 1{\cdot}6$ °C ✓

(to 1 dec pl.)

4 marks

Mean

- The symbol $\overline{x}$ ("x bar") is used for the mean
- Mean $= \frac{\text{sum of the numbers}}{\text{number of numbers}} = \frac{\Sigma x}{n}$

Squared Deviations

- This mark is for correctly calculating the squared deviations from the mean $(x-\overline{x})^2$. i.e. getting the last column of the table correct.

Substitution

- Correct substitution of $\Sigma(x-\overline{x})^2 = 16$ and $n-1=6$ into the formula will gain this mark

Calculation

- Don't forget to take the square root and to round your answer
- The alternative formula is

$$s = \sqrt{\frac{\Sigma x^2 - (\Sigma x)^2/_n}{n-1}}$$

In this case $\Sigma x^2 = 863$, $(\Sigma x)^2 = 5929$, $n = 7$ and $n-1=6$ giving :

$$s = \sqrt{\frac{863 - 5929/_7}{6}} = \sqrt{\frac{16}{6}} \doteq 1{\cdot}6 \text{ as before}$$

- Both formulae are given on your formulae page during the exam. The formula used in the solution opposite is usually easier to use with less possibility of making mistakes. However you should use the formula you are most comfortable using.

See Int 2 Notes, Section 9, p. 33–34

Q 4. *(b)*

New mean = 9°C ✓

New standard deviation = 1·6°C ✓

2 marks

Mean

- All 7 numbers are reduced by 2. The total reduces by 14. Dividing by 7, therefore reduces the mean by 2

Standard Deviation

- The distribution about the mean will be unaltered so the standard deviation will remain the same

Q 5. *(a)*

$3x + 5y = 190$ ✓

1 mark

1st equation

- Interpretation is a particular difficulty in these 'simultaneous equation' questions. There is a lot of writing to interpret and to extract the necessary information from. To help you, remember there will always be two quantities that are not known. In this case the two different fees. Concentrate on one of these, say £x, the fee for the beginners class. What information is given? "3 people turned up" So 3 fees of £x giving £$3x$. Similarly for the advanced class: "5 people turned up", so 5 fees of £y giving £$5y$. In total the fees are £$(3x + 5y)$. But you are told the total fees were £190 so $3x + 5y = 190$.

Q 5. *(b)*

$4x + 2y = 132$ ✓

1 mark

2nd equation

- Be careful not to mix up the xs and ys! This is a common mistake leading to the wrong equation: $2x + 4y = 132$. Your only consolation is that in part(c) you can still achieve full marks using the wrong equation

Q 5. *(c)*

Solve:

$$\left.\begin{aligned}3x+5y&=190\\4x+2y&=132\end{aligned}\right\}\begin{aligned}&\times 4 \rightarrow 12x+20y=760\\&\times 3 \rightarrow \underline{12x+6y=396}\end{aligned}$$

Subtract: $14y = 364$

$\Rightarrow y = \frac{364}{14} = 26$ ✓

Substitute $y = 26$ in $3x + 5y = 190$

$\Rightarrow 3x + 5 \times 26 = 190$

$\Rightarrow 3x + 130 = 190$ ✓

$\Rightarrow 3x = 190 - 130 = 60$

$\Rightarrow x = \frac{60}{3} = 20$ ✓

So the 'beginner's class' costs £20 and the 'advanced class' costs £26. ✓

4 marks

Strategy

- This mark is gained for evidence that you know this is a 'simultaneous equation' question. You will gain this mark for just 'starting' the process.

Values

- Working through to a pair of values – even if they are wrong will gain you this mark.

Correct values

- The alternative method is to find x first:

$$\left.\begin{aligned}3x+5y&=190\\4x+2y&=132\end{aligned}\right\}\begin{aligned}&\times 2 \rightarrow 6x+10y=380-(1)\\&\times 5 \rightarrow \underline{20x+10y=660}-(2)\end{aligned}$$

subtract (1) from (2): $14x = 280$

$\Rightarrow x = \frac{280}{14} = 20$

Then substitute in one of the two equations

- Always check, in this case use $x = 20$, $y = 26$ in $4x + 2y$ to give $4 \times 20 + 2 \times 26 = 80 + 52 = 132$

Statement

- Fees were asked for so your final statement must answer the question.

See Int 2 Notes, Section 7, p. 26–27.

Q 6.

Miles travelled = 2500

So Chargeable miles = 2500 – 1000
= 1500 ✓

$\Rightarrow$ Chargeable miles > 0

So Distance charge = 1500 × 19·5p
= 29250p
= £292·50

Number of days = 5
which is not greater than 5 ✓

So

Hire charge = 5 × £25 + £292·50
= £125 + £292·50
= £417·50 ✓

3 marks

Chargeable Miles
- The tophalf of the flowchart is concerned with the number of miles the hired car has travelled namely 2500. So the first decision box has a 'YES' answer

Correct Path
- YES followed by NO gains you this mark.
- Note that $x > 5$ in whole numbers has solution: 6, 7, 8,... 5 > 5 is not true. This means a 'NO' answer for the 2nd decision box.
- Notice that in many flowcharts a quality is named and calculated and then used later. In this case the quality 'Chargeable miles' and also 'Distance charge'

Calculation
- The formula:
  Hire charge = Number of days × £25 + Distance charge is used with:
  Number of days = 5
  and Distance charge = £292·50

Q 7.

Find the two volumes ✓

The 'wok'

Volume of a sphere = $\frac{4}{3}\pi r^3$

So Volume of a hemisphere
$= \frac{1}{2} \times \frac{4}{3}\pi r^3 = \frac{2}{3}\pi r^3$

In this case $r = \frac{31}{2}$ = 15·5 cm ✓

So Volume = $\frac{2}{3} \times \pi \times 15{\cdot}5^3$
= 7799·26.... ≑ 7800 cm^3
(to 3 sig figs)

The 'pan'

Volume of a cylinder = $\pi r^2 h$

In this case r = 25cm and
h = 3·5cm

So Volume = $\pi \times 25^2 \times 3{\cdot}5$ ✓
= 6872·23.... ≑ 6900 cm^3
(to 3 sig figs) ✓

So the 'wok' has the larger capacity by approximately 900 cm^3 ✓

5 marks

Strategy
- Capacity' means volume. You will have to calculate the volume of each type of pan and compare these volumes.

1st substitution
- The formula V = $\frac{4}{3}\pi r^3$ is given to you on the formulae page during your exam
- In this case you have only half a sphere so use $\frac{1}{2}$ of $\frac{4}{3}\pi r^3$ i.e $\frac{2}{3}\pi r^3$ with r being half of the diameter of 31cm i.e. 15·5 cm.

2nd substitution
- The formula V = $\pi r^2 h$ is given on your formulae sheet
- In this case substitute r = 25 and h = 3·5

Volumes
- Both volumes correct and you gain this mark

Statement
- You must clearly justify your conclusion using the values for the volume. Finding the difference and stating shows clearly that you understand the results you calculated.

Q 8_(a)_

Use the Cosine Rule in triangle ABC: ✓

$a^2 = b^2 + d^2 - 2bd \cos A$

$= 6{\cdot}3^2 + 4{\cdot}5^2 - 2 \times 6{\cdot}3 \times 4{\cdot}5 \cos 79°$

$= 49{\cdot}12....$ ✓

So $a = \sqrt{49{\cdot}12....}$

$= 7{\cdot}008....$

$\doteqdot 7{\cdot}01$ cm ✓

The distance between vertex B and vertex D is approximately 7·01cm

(Correct to 3 significant figures)

3 marks

Strategy

- You gain this mark if there is evidence in your solution that you knew to use the Cosine Rule.
- How do you know to use the Cosine Rule and not the Sine Rule? In this case you know side AB and side AD and the angle inbetween (the 'included' angle) angle BAD.

  To use the Cosine Rule you will always need to know two sides and the included angle (if you are finding the 3rd side)

Substitution

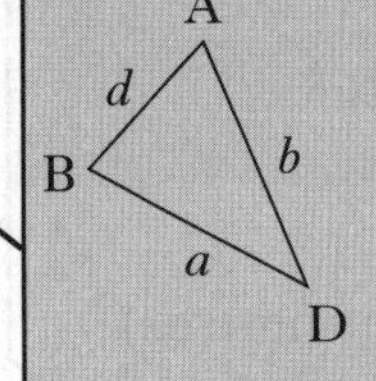

- Your formulae page gives:

  $a^2 = b^2 + c^2 - 2bc \cos A$

  You have to know how to adapt this for a triangle named ABD:

  $a^2 = b^2 + d^2 - 2bd \cos A$

where $b = 6{\cdot}3$, $d = 4{\cdot}5$ and angle A = 79°

Calculation

- Remember to find the square root as the last step in the calculation
- You should estimate from the diagram roughly the length you would expect for the answer. If your calculated answer is not near this estimate then check your calculation.

  From the diagram BD looks to be slightly longer than AD = 6·3 cm so around 7 cm is a reasonable length.

  See Int 2 Notes, Section 6, p. 24

Q 8. *(b)*

Required Area = Area of triangle ABD − Area of triangle BCD ✓

Area of triangle BCD

$= \frac{1}{2} \times \text{BC} \times \text{BD} \times \sin \text{C}\hat{\text{B}}\text{D}$

$= \frac{1}{2} \times 3{\cdot}6 \times 7{\cdot}008\ldots \times \sin 15°$

$= 3{\cdot}265\ldots \text{ cm}^2$ ✓

Area of triangle ABD

$= \frac{1}{2} \times \text{AB} \times \text{AD} \times \sin \text{B}\hat{\text{A}}\text{D}$

$= \frac{1}{2} \times 4{\cdot}5 \times 6{\cdot}3\ldots \times \sin 79°$

$= 13{\cdot}914\ldots \text{ cm}^2$ ✓

Area of plate

$= 13{\cdot}914\ldots - 3{\cdot}265\ldots$

$= 10{\cdot}649\ldots$

$\doteqdot 10{\cdot}6 \text{ cm}^2$ ✓

(correct to 3 significant figures)

4 marks

Strategy
- A subtraction of two areas is the plan.

1st substitution
- You use the area formula: Area = $\frac{1}{2}$ $ab \sin C$ which is given on your formulae sheet.
- Do not use the rounded version from part(a) for the length of BD.

2nd substitution
- Using $\frac{1}{2} bd \sin A$ with $b = 6{\cdot}3$, $d = 4{\cdot}5$ and angle A = 79°

Calculation
- Again do not round answers before the end of the calculation as you will then be introducing errors

See Int 2 Notes, Section 6, p. 22

Q 9. *(a)*

£111·80 ✓

1 mark

Repayment
- 5 years is 5 × 12 = 60 months.
- Be careful to choose the 'without payment protection' column — the left hand column.

Q 9. *(b)*

£111·80 × 60
= £6708·00 ✓

1 mark

Total
- He pays 60 repayments of £111·80 over the 5 years of the loan. So he pays back a total of £6708·00 for this loan.

Q 9. *(c)*

Cost of the £15000 loan
= £205·64 × 120
= £24676·80

Total cost of two separate loans
= £6708·00 + £24 676·80
= £31384·80 ✓

Cost of the single £20 000 loan
= £274·18 × 120
= £32 901·60

The single £20 000 is more expensive (by £1516·80) ✓

2 marks

Two loans
- The 2nd loan is paid back over a longer term i.e. 10 years or 10 × 12 = 120 months.
- The strategy is to calculate the cost of the two separate loans (one for 5 yrs and one for 10 yrs) and then add these costs together.

Comparison
- The total cost of two separate loans has to be compared with the total cost of the one larger £20000 loan.
- A comparison should be clear that explains your answer — the difference in cost is the ideal way of comparing them.

Q 10.

Score	(f) Frequency	(x) Mid-point	(fx) Mid-point × frequency
0–6	1	3	3
7–13	4	10	40 ✓
14–20	11	17	187 ✓
21–27	8	24	192
28–34	3	31	93
	$\sum f = 27$		$\sum fx = 515$ ✓

$$\text{Mean} = \frac{\sum fx}{\sum f} = \frac{515}{27} = 19{\cdot}07\ldots \quad ✓$$

$$\doteqdot 19{\cdot}1 \quad ✓$$

5 marks

Midpoints

- It is assumed that all the scores in an interval lie at the mid point. So for 7–13 the mid point is $\frac{7+13}{2} = 10$ and it is assumed these were 4 scores of 10 marks.
- Check the pattern of mid points: $3 + 7 = 10$, $10 + 7 = 17$, $17 + 7 = 24$ and $24 + 7 = 31$.

fx

- This mark is for the correct last column.

Totals

- Correct calculation of $\sum f$ and $\sum fx$ gains this mark.

Mean

- You should know: Mean $= \frac{\sum fx}{\sum f}$.

Calculation

- Any correct rounding is allowed for the final answer.

Q 1. *(a)*
$3k + (2k - 3)(k - 4)$ ✓
$= 3k + 2k^2 - 8k - 3k + 12$ ✓
$= 2k^2 - 8k + 12$ ✓

3 marks

Start brackets
- You should use this pattern:

Firsts Lasts
$(2k - 3)(k - 4)$
Insides
Outsides

to give (FOIL):

$2k \times k$	$2k \times (-4)$	$-3 \times k$	$-3 \times (-4)$
(Firsts)	(Outsides)	(Insides)	(Lasts)
$2k^2$	$-8k$	$-3k$	12

Two terms correct gains you the 1st mark

Complete brackets
- All 4 terms correct from multiplying out the brackets gains you this 2nd mark

Like terms
- You must 'tidy up' gathering all the like terms together and simplifying:

$3k - 8k - 3k = - 8k$

See Int 2 Notes, Section 4, p. 12–14

Q1. *(b)*
$7m^2 + 54m - 16$ ✓
$= (7m - 2)(m + 8)$ ✓

2 marks

Start to factorise
- $7m^2$ gives only one possibility namely (7m ....)(m ....)
- 16 has many possible factors:

(....4)(....4) or (....2)(....8) etc

Finish factoring
- When you think you have the correct combination then use "FOIL" to multiply the brackets out. If your answer is not $7m^2 + 54m - 16$ then try another combination!

See Int 2 Notes, Section 4, p. 16

Q 2.
Probability $= \frac{4}{16} = \frac{1}{4}$ ✓

1 mark

Probability

- You have to identify those passengers who waited for more than 15 minutes. How many are there. Fortunately a stem-and-leaf diagram shows the numbers in increasing order.

  15 minutes would show up between the '2' and '8' in the 2nd row of the diagram. There are only 4 times greater than 15 namely:

  18, 21, 26 and 30

- 4 passengers out of a total of 16 passengers gives a probability of $^4/_{16}$ or $^1/_4$.

See Int 2 Notes, Section 9, p. 35–36

Q 3. *(a)*(i)
Median = 110 minutes ✓

1 mark

Median (Q_2)

- Read the stem-and-leaf diagram left to right, top to bottom: this gives the data in increasing order:

  (85 93 101 102 107 109) 110 (113 114 117 119 119 120)

  The median is the middle value

Q3. *(a)*(ii)
Lower quartile $= \frac{101+102}{2} = 101{\cdot}5$ ✓
Upper quartile $= \frac{117+119}{2} = 118$ ✓

2 marks

Lower quartile (Q_1)

- This is the median of the lower 6 values i.e. The mean of the two middle values 101 and 102

Upper quartile (Q_3)

- This is the median of the upper 6 values i.e. The mean of the two middle values 117 and 119

See Int 2 Notes, Section 8, p. 31

Q 3. *(b)*

Boxplot of walk times ✓ ✓

85 90 95 100 105 110 115 120
Minutes

2 marks

End points

- You should first find the least and greatest values: in this case 85 and 120. This allows you to construct the scale
- Use the squared paper provided in your exam

Box

- Q_1 and Q_3 mark the ends of the box, in this case the values are 101·5 and 118
- Q_2 is shown by the line in the box, in this case the value is 110

See Int 2 Notes, Section 8, p. 31

Q 4. *(a)*

gradient $= -\frac{1}{2}$ ✓

1 mark

Gradient

- Gradient $= \frac{\text{Distance up (or down)}}{\text{Distance along}}$. In this case the line goes 1 down, 2 along. 1 down is shown as -1 in the fraction: $^{-1}/_2$ giving $-^1/_2$

See Int 2 Notes, Section 3, p. 9

Q 4. *(b)*

Equation is: $y = -\frac{1}{2}x + 2$ ✓

✓

$(\Rightarrow 2y = -x + 4 \Rightarrow 2y + x = 4)$

2 marks

***y*-intercept**

- The y-intercept is (0, 2) giving $c = 2$

Equation

- You use '$y = mx + c$' with $m = -\frac{1}{2}$ and $c = 2$
- Further rearrangement is not essential

See Int 2 Notes, Section 3, p. 10–11

Q 4. *(c)*

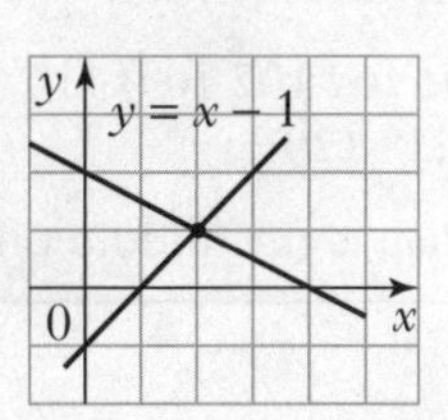

✓

The point of intersection is (2,1) ✓

2 marks

2nd line

- $y = x - 1$ has gradient 1 (1 up, 1 along) and y-intercept (0, − 1)

Intersection

- This can be read off the graph: the point where the two lines cross i.e. (2,1)
- Alternatively solve simultaneously:

$$\left.\begin{aligned} y &= -\tfrac{1}{2}x + 2 \\ y &= x - 1 \end{aligned}\right\} \begin{aligned} &\rightarrow 2y + x = 4 \\ &\rightarrow y - x = -1 \end{aligned}$$

Add: $3y = 3 \Rightarrow y = 1$ etc

See Int 2 Notes, Section 3, p. 10

Q 5.

$\cos x^\circ = 0{\cdot}951$

x° is in the 1st or 4th quadrants

The 1st quadrant angle is 18°

The 4th quadrant angle is

$360^\circ - 18^\circ = 342^\circ$

So $x = 342$ is therefore another value ✓

1 mark

Value

- You have no calculator in paper 1. The information given in the question is the following: $\cos^{-1}(0{\cdot}951) = 18^\circ$ which is what you normally get from your calculator
- $\cos x^\circ$ is positive. Use:

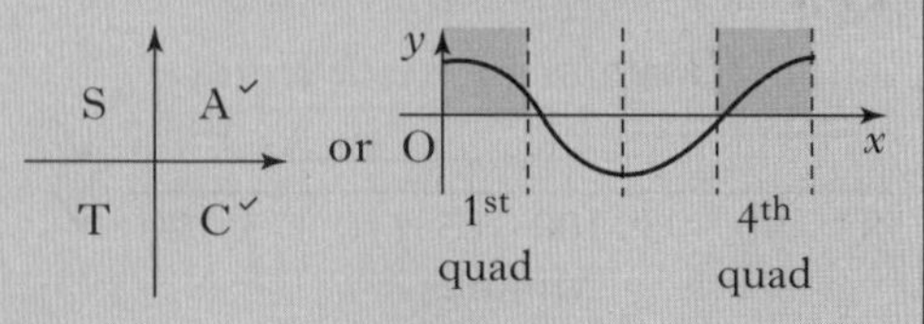

- The 4th quadrant angle is always given by:

$$360^\circ - (1^{\text{st}}\text{ quadrant angle})$$

See Int 2 Notes, Section 12, p. 50

Q 6.
Yes it is Hamiltonian ✓
The following is one possible route that shows this:
$A \to E \to B \to D \to C \to F \to A$ ✓

2 marks

Statement
- Your statement 'Yes' must be accompanied by a reason.

Reason
- In this case a route that passes through all of A, B, C, D, E and F *and* returns to where it started should be given
- There are many such routes

Q 7. *(a)*
£362·90 ✓

1 mark

Amount
- This formula adds the amounts appearing in cells D2 and F2, namely £268·90 and £94. The result £362·90 appears in cell G2

Q 7. *(b)*
= B3 * 2 * E3 ✓

1 mark

Formula
- '=' is not essential to gain the marks
- × can be used instead of * to gain the mark
- order is not important eg 2 × B3 × E3 is ok

Q 8. *(a)*
$l = 12,\ k = 0{\cdot}004,\ t = 5$
$L = l(1 + kt)$
$\Rightarrow L = 12 \times (1 + 0{\cdot}004 \times 5)$ ✓
$= 12 \times (1 + 0{\cdot}02)$ ✓
$= 12 \times 1{\cdot}02 = 12{\cdot}24$ ✓

3 marks

Strategy
- You gain this mark from working that shows you know to substitute the values.

Substitution
- Correct replacement of each letter with its corresponding value will gain you this work

Calculation
- You have no calculator in Paper 1. You should therefore look for 'easier' ways of calculating.
  For example 12 × 1·02. Calculate 10 × 1·02.
  Then 2 × 1·02 and add the results

Q 8. *(b)*

$l = 5$, $k = 0{\cdot}007$, $L = 5{\cdot}28$

$L = l(1 + kt)$

$\Rightarrow 5{\cdot}28 = 5 \times (1 + 0{\cdot}007t)$ ✓

$\Rightarrow 5{\cdot}28 = 5 + 0{\cdot}035t$ ✓

$\Rightarrow 0{\cdot}28 = 0{\cdot}035t$

$\Rightarrow \dfrac{0{\cdot}28}{0{\cdot}035} = t$

$\Rightarrow t = \dfrac{280}{35} = 8$ ✓

3 marks

Substitution
- This first mark is for correctly replacing each letter with its corresponding value

Equation
- Starting to solve the resulting equation will earn you the second mark.

Solution
- Subtract 5 from both sides of the equation and then divide both sides by $0{\cdot}035$
- Notice $\dfrac{0{\cdot}28}{0{\cdot}035} = \dfrac{0{\cdot}28 \times 1000}{0{\cdot}035 \times 1000} = \dfrac{280}{35} = \dfrac{7 \times 40}{7 \times 5}$

Q 9. *(a)*

Area = length × breadth

$= (x + 1)(x - 2)\text{ cm}^2$ ✓

1 mark

Area expression
- You are using the area of a rectangle formula here

Q 9. *(b)*

Frame Area = Area of outside Rectangle − Area of Inside Rectangle ✓

$= x(x + 3) - (x + 1)(x - 2)$

$= x^2 + 3x - (x^2 - 2x + x - 2)$

$= x^2 + 3x - x^2 + 2x - x + 2$

$= 4x + 2\text{ cm}^2$ ✓

2 marks

Strategy
- The frame is what is left when you remove the picture. This leads to the strategy of subtraction of the picture area (from part (a)) from the large rectangle area ($x \times (x + 3)$)

Simplification
- Brackets are essential in $(x^2 - 2x + x - 2)$ since you are subtracting all of this expression.
- The 'x^2' terms cancel and the like terms $3x$, $2x$ and $-x$ simplify to $4x$

See Int 2 Notes, Section 4, p. 12–14

WORKED ANSWERS: EXAM C — PAPER 2

Q 1.
Commission = $1{\cdot}2\%$ of £180500
= £2166 ✓
Cost (before VAT)
= £2166 + £325 = £2491 ✓

2 marks

Strategy
- The VAT rate of $17{\cdot}5\%$ is not used to answer this question
- Cost = commission + fee (before VAT).

Calculation
- On the calculator to find $1{\cdot}2\%$: $1{\cdot}2 \div 100 \times 180500$
- You would still gain the marks if you went on and added the VAT but that extra calculation is not needed and could waste your time in the exam

Q 2.
Let size A weigh a kg and size B weigh b kg.
The 1st diagram gives:
$3a + 2b = 94$ ✓
The 2nd diagram gives:
$2a + 3b = 101$ ✓
Solve:

$$\left.\begin{aligned}3a+2b&=94\\2a+3b&=101\end{aligned}\right\}\begin{aligned}&\times 3 \rightarrow 9a+6b=282\\&\times 2 \rightarrow 4a+6b=202\end{aligned}$$ ✓

Subtract : $5a = 80$

$\Rightarrow a = \frac{80}{5} = 16$

Substitute $a = 16$ in $3a + 2b = 94$
$\Rightarrow 3 \times 16 + 2b = 94$
$\Rightarrow 48 + 2b = 94$ ✓
$\Rightarrow 2b = 94 - 48 = 46$
$\Rightarrow b = 23$ ✓
One size A weight weighs 16 kg and one size B weight weighs 23 kg ✓

6 marks

Interpretation
- Make sure the two quantities you are to find are given letters and explain what the two letters are standing for
- Three Size A weights will weigh $3 \times a$ kg and two size B weights will weigh $2 \times b$ kg.

2nd equation
- Two Size A and three Size B weigh $2 \times a + 3 \times b$ kg

Strategy
- Two equations with two unknowns: you need to use simultaneous equations

Two values
- Following the method to produce two values will gain you this mark - even for the wrong values!

Correct values
- You should check the values work in the 'other' equation i.e. $2a + 3b = 101$

Statement
- Reread the question: what exactly did it ask you to find? Make sure you have done this!

See Int 2 Notes, Section 7, p. 25–27.

Q 3.
A reduction of 22% means that 78% will be left giving a multiplication factor of 0·78 ✓
After 3 minutes:
N° of bacteria = $250000 \times 0{\cdot}78^3$ ✓
= 118638 ≑ 119000
(to the nearest thousand) ✓

3 marks

Factor
- Notice: $78\% = \frac{78}{100} = 0{\cdot}78$

Strategy
- To find 78% of a quantity multiply by 0·78. Doing this 3 times ($0{\cdot}78 \times 0{\cdot}78 \times 0{\cdot}78$) is equivalent to multiplying by $0{\cdot}78^3$

Calculation
- On your calculator $\times\, 0{\cdot}78^3$ is keyed as: [×][0][·][7][8][^][3]. The key [^] means "raised to the power of".
- Any correct rounding is acceptable (or no rounding!)

See Int 2 Notes, Section 1, p. 5

Q 4. *(a)*

x	$x - \bar{x}$	$(x - \bar{x})^2$
503	2	4
504	3	9
497	–4	16
495	–6	36
506	5	25

✓

$\Sigma(x - \bar{x})^2 = 90$

$S = \sqrt{\frac{\Sigma(x-\bar{x})^2}{n-1}} = \sqrt{\frac{90}{4}}$ ✓
$= 4{\cdot}743\ldots \doteqdot 4{\cdot}7$ ✓
(to 1 decimal place)

3 marks

Squared Deviations
- This mark is gained for correctly calculating the values in the last column of the table i.e $(x - \bar{x})^2$
- No negative values should ever appear in this column. Remember when you square a quantity your answer will be positive or zero
- The mean, $\bar{x}$, has been given to you in the question and so you do not need to calculate it again!

Substitution
- This mark is for correct substitution of the values $\Sigma(x - \bar{x})^2 = 90$ and $n - 1 = 4$ into the standard deviation formula

Calculation
- Remember always that there is a square root to be taken at the end of a standard deviation calculation
- Any reasonable rounding is acceptable

See Int 2 Notes, Section 9, p. 33–34

Q 4. *(b)*

Yes it did. The new standard deviation of 3·5 is less than the previous value of 4·7 so there was less variation about the mean. ✓

1 mark

Statement

- Your explanation must use the standard deviation statistics that you calculated in part(a) and are also given in the question. The greater the standard deviation the greater the variation about the mean (and vice versa).

Q 5.

Fraction of circle $= \frac{40}{360} = \frac{1}{9}$ ✓

So Shaded Area $= \frac{1}{9} \times \pi r^2$ ✓

where $r = \frac{2 \cdot 5}{2} = 1 \cdot 25$ metres

Shaded Area $= \frac{1}{9} \times \pi \times 1 \cdot 25^2$

$= 0 \cdot 5454....$

$\doteqdot 0 \cdot 545 \text{ m}^2$ ✓

(correct to 3 significant figures)

3 marks

Fraction

- The sector shown has a central angle of 40°. The complete circle has a central angle of 360°. Comparing 40° with 360° gives the fraction $\frac{40}{360}$
- Cancelling down is not essential but is neater.

Strategy

- You have a sector that is $\frac{1}{9}$ of the complete circle and it will therefore have an area that is $\frac{1}{9}$ of πr^2. The diameter is given as 2·5 metres. Remember to half this before substitution into πr^2.

Calculation

- Always use the $\boxed{\pi}$ button on your calculator, not values like 3·14 unless the question asks you to do this
- Accuracy is not mentioned in the question so there is no mark awarded for rounding

See Int 2 Notes, Section 5, p. 17–18

Q 6.

For y-axis intercept

set $x = 0$ in $2x - 3y = 12$ ✓

$\Rightarrow 2 \times 0 - 3y = 12$

$\Rightarrow 0 - 3y = 12$

$\Rightarrow -3y = 12$

$\Rightarrow y = \frac{12}{-3} = -4$

Required point is (0, –4) ✓

2 marks

Strategy

- On the y-axis all points have x-coordinate equal to zero i.e. $x = 0$. All points on the line have x and y-coordinates that satisfy $2x - 3y = 12$. You therefore need to substitute $x = 0$ in the equation to find the point on the line that is also on the y-axis.
- Alternatively you could draw a graph of the line to see where it crosses the y-axis

Point

- $y = -4$ is not sufficient. A point is required. So coordinates need to be shown as: (0, –4)

Q 7.

At 10·4% over 180 months a £5000 loan costs £53·64 per month: ✓

Amount		Cost
£5000	:	£53·64
£5000	:	£53·64
£5000	:	£53·64
£2500	:	£26·82
£17500	:	£187·74 ✓

Total cost = £187·74 × 180 ✓

= £33 793·20 ✓

4 marks

Table

- The 1st mark is for interpreting the table correctly to get £53·64 per month

Strategy

- You should realise that £17500 can be split into 3 lots of £5000 and one half of £5000 to give 3½ × £53·64

Strategy

- You should know to multiply the total monthly repayment by 180 for the total cost
- Strategy marks can still be gained if you use the wrong table entry.

Calculation

- You gain the final mark for a correct calculation.
- The cost is large compared to the loan. This is because it is paid back over a long term namely 15 years

Q 8. *(a)*

Extend line AB (see diagram)

$\angle NBE = 105°$

So $\angle CBE = 205° - 105°$

$= 100°$

So $\angle ABC = 180° - 100°$

$= 80°$ ✓

(since $A\hat{B}E$ is a straight angle)

1 mark

Angle

- It is essential to bring the 105° information at vertex A to vertex B. This can be done by extending the line AB
- An alternative is:

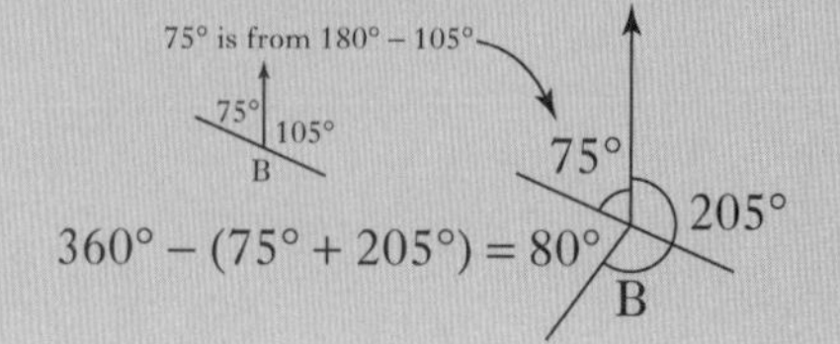

Q 8. *(b)*

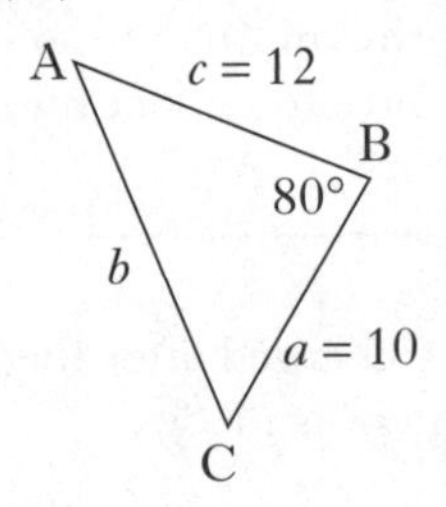

Use the Cosine Rule in triangle ABC: ✓

$b^2 = a^2 + c^2 - 2ac \cos B$

$= 10^2 + 12^2 - 2 \times 10 \times 12 \times \cos 80°$

$= 202{\cdot}32....$ ✓

So $b = \sqrt{202{\cdot}32....}$

$= 14{\cdot}22... \doteqdot 14{\cdot}2$ km ✓

The required distance is 14·2 km (to 1 dec pl)

3 marks

Strategy
- Two sides and the included angle all known means the Cosine Rule should be used

Substitution
- The version given on your formulae page is:

$$a^2 = b^2 + c^2 - 2bc \cos A$$

You must be able to change this to find side b:

$$b^2 = a^2 + c^2 - 2ac \cos B$$

Practice this!

Calculation
- Remember the square root at the end.

See Int 2 Notes, Section 6, p. 24

Q 8. *(c)*

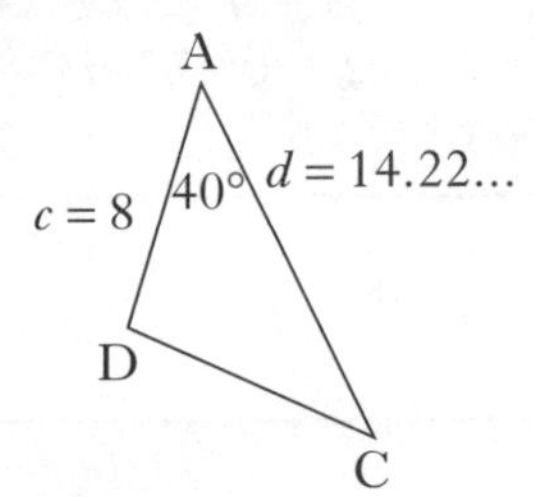

Area of triangle ADC

$= \frac{1}{2} cd \sin A$ ✓

$= \frac{1}{2} \times 8 \times 14{\cdot}22... \times \sin 40°$

$= 36{\cdot}57....$

$\doteqdot 36{\cdot}6$

Required area is 36·6 km^2 (correct to 1 decimal place) ✓

2 marks

Strategy
- This is not a right-angled triangle so the 'trig' area formula will have to be used. The formula given in your exam is:

Area = $\frac{1}{2} ab \sin C$

You should adapt this for triangle ACD:

Area = $\frac{1}{2} cd \sin A$

- The formula always involves two sides and the angle in between them (the included angle).

Calculation
- Do not use rounded answers in subsequent calculations. In this case using the rounded answer 14·2 km from part (*b*) gives an answer of 36·5 km^2. You cannot say this is accurate to 1 decimal place!

See Int 2 Notes, Section 6, p. 22

Q 9. *(a)*

Total Due = £318·70

So minimum monthly payment

= 3% of £318·70 ✓

= £9·56 (to the nearest penny) ✓

2 marks

Interpretation
- You have to read carefully all the information written in the Credit Card Statement. "Minimum monthly payment is 3% of the Total Due". The Total Due is stated to be £318·70

Calculation
- The calculator calculation is: $3 \div 100 \times 318{\cdot}7$ giving 9·561 which is rounded to £9·56

Q 9. *(b)*
Amount due after monthly payment
= £318·70 – £9·56 ✓
= £309·14
Interest for month
= 2·2% of £309·14 ✓
= £6·80 (to the nearest penny)
Total due after interest and purchases:
= £309·14 + £6·80 + £150
= £465·94 ✓

3 marks

Strategy
- Once you pay the minimum monthly payment you have to deduct this from the amount due

Strategy
- Interest is now charged after the monthly payment is made
- You should round answers to the nearest penny in questions like this

Calculation
- The total due consists of three items:
  - The previous month's amount due
  - The month's interest
  - The month's purchases

Q 10. *(a)*
The entries in the last column are:
1, 4, 12, 29, 41, 50, 57, 62, 63, 64 ✓

1 mark

Cumulative frequencies
- An extra column should be added to the table and the 'running totals' entered:

  1 + 3 = 4, 4 + 8 = 12, 12 + 17 = 29, 29 + 12 = 41, 41 + 9 = 50, 50 + 7 = 57, 57 + 5 = 62, 62 + 1 = 63 and 63 + 1 = 64.

Q 10. *(b)*

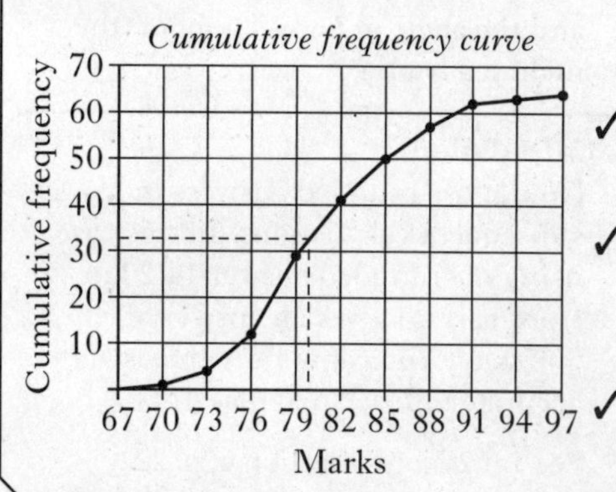

✓ ✓ ✓

3 marks

Scales
- Choose scales that allow a reasonable size for your diagram. The numbers along the 'x-axis' are the upper bound of each class interval eg 68–70 uses 70 etc

Points
- Plot the points (67, 0), (70, 1), (73, 4), (76, 12), (79, 29) etc to gain this mark

Curve
- Join the points, in pairs, to produce the cumulative frequency curve as shown.

Q 10. *(c)*
The median is approximately 80. ✓

1 mark

Median
- The total of the frequencies is 64. Half this to get 32. Draw a horizontal line from 32 on the vertical axis to meet the curve. Now drop a vertical line from this meeting place and read off the median on the horizontal axis.

Q 11. *(a)*

Volume of cone = $\frac{1}{3}\pi r^2 h$

with $r = \frac{4}{2} = 2$ cm and $h = 6$ cm

So Volume = $\frac{1}{3} \times \pi \times 2^2 \times 6$ ✓

= 25·132.... ✓

≑ 25·1 cm^3

(to 1 dec pl.) ✓

3 marks

Substitution

- This 1st mark is for correctly substituting the values $r = 2$ cm and $h = 6$ cm into the Cone formula
- The formula is given to you on your formulae page during your exam
- The radius (r) is used in the formula. You are given the diameter (4 cm) and so need to halve this

Calculation

- Remember to use the $\boxed{\pi}$ button (not 3·14)

Rounding

- There is a mark allocated for correct rounding to 3 significant figures. An answer of 25 cm^3 would not gain this mark – it's only got 2 sig figs!

See Int 2 Notes, Section 2, p. 8

Q 11. *(b)*

Volume of a cylinder = $\pi r^2 h$

So Volume of the weight

= $\frac{1}{2}\pi r^2 h$

where

$r = 2$ cm, Volume = 25·132... cm^3

and h is not known

$\Rightarrow 25{\cdot}132... = \frac{1}{2} \times \pi \times 2^2 \times h$ ✓

$\Rightarrow 25{\cdot}132... = 2\pi \times h$ ✓

$\Rightarrow h = \frac{25{\cdot}132....}{2\pi} = 4$ ✓

The height of the weight is 4 cm

3 marks

Volume of weight

- Correct substitution in $\frac{1}{2}\pi r^2 h$ gains this mark

Equation

- Setting the volume of the weight equal to the volume of the cone calculated in part (a) is the strategy for finding h.
- Don't use 25·1 cm^3 – This was rounded.

Solving

- Solve the equation to gain this mark
- You might wonder at exactly 4 cm being the solution. Is it exactly 4? Yes it is:

  $\frac{1}{3}\pi \times 2^2 \times 6 = \frac{1}{2}\pi \times 2^2 \times h$

  (divide by $\pi \times 2^2$)

  $\Rightarrow \frac{1}{3} \times 6 = \frac{1}{2}h \Rightarrow 2 = \frac{1}{2}h \Rightarrow h = 4$

  This proof uses the 'exact' expression for the volume of the cone from part (a).

See Int 2 Notes, Section 2, p. 8

Q 12. *(a)*

$\angle AOB = \left(\frac{360}{5}\right)^\circ = 72^\circ$ ✓

1 mark

Angle

- There would be 5 equal angles around O if the other radii were drawn in. So each angle is $\frac{1}{5}$ of 360°.

Q 12. *(b)*

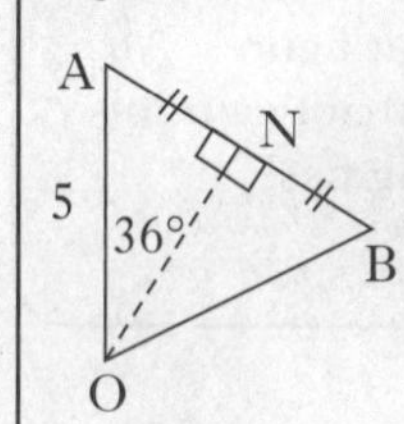

Draw ON, the altitude of triangle AOB

ON bisects angle AOB so Angle AON = 36° ✓

In triangle AON, $\sin 36^\circ = \frac{AN}{5}$

$\Rightarrow AN = 5 \sin 36^\circ = 2{\cdot}938\ldots$ ✓

So $AB = 2 \times AN = 2 \times 2{\cdot}938\ldots$

$= 5{\cdot}877\ldots$

So AB ≑ 5·88 cm ✓

(to 3 significant figures)

3 marks

Strategy

- Create a right-angled triangle and use 'SOHCAHTOA'
- Alternatively you could use the Sine Rule in triangle AOB (The angles are 72°, 54° and 54°)

SOHCAHTOA

- In triangle OAN you are trying to find the 'Opposite' and you know the 'Hypotenuse' This gives (SOH)CAHTOA so use 'sin'

Side length

- By the symmetry of the diagram, doubling your answer for the length AN will give the length AB

See Int 2 Notes, Section 5, p. 20
Section 6, p. 21